T0181003

Springer Series on Naval Architecture, Marine Engineering, Shipbuilding and Shipping

Volume 12

Series Editor

Nikolas I. Xiros, University of New Orleans, New Orleans, LA, USA

The Naval Architecture, Marine Engineering, Shipbuilding and Shipping (NAMESS) series publishes state-of-art research and applications in the fields of design, construction, maintenance and operation of marine vessels and structures. The series publishes monographs, edited books, as well as selected PhD theses and conference proceedings focusing on all theoretical and technical aspects of naval architecture (including naval hydrodynamics, ship design, shipbuilding, shipyards, traditional and non-motorized vessels), marine engineering (including ship propulsion, electric power shipboard, ancillary machinery, marine engines and gas turbines, control systems, unmanned surface and underwater marine vehicles) and shipping (including transport logistics, route-planning as well as legislative and economical aspects).

The books of the series are submitted for indexing to Web of Science.

All books published in the series are submitted for consideration in Web of Science.

More information about this series at https://link.springer.com/bookseries/10523

Arun Kr Dev · Makaraksha Saha · George Bruce

Ship Repairing

Analyses and Estimates

 Springer

Arun Kr Dev
Newcastle University
Singapore, Singapore

Makaraksha Saha
Independent Marine Consultant
Singapore, Singapore

George Bruce
Newcastle University
Newcastle upon Tyne, UK

ISSN 2194-8445   ISSN 2194-8453 (electronic)
Springer Series on Naval Architecture, Marine Engineering, Shipbuilding and Shipping
ISBN 978-981-16-9470-7   ISBN 978-981-16-9468-4 (eBook)
https://doi.org/10.1007/978-981-16-9468-4

This Springer imprint is published by the registered company Springer Nature Singapore Pte Ltd.
The registered company address is: 152 Beach Road, #21-01/04 Gateway East, Singapore 189721,
Singapore

Foreword by Prof. Choo Chiau Beng

When I joined the ship repair industry in 1971 upon returning to Singapore after my MSc in Naval Architecture from Newcastle, there were hardly any reading materials on ship repair.

Somehow academics did not feel it was a strong enough subject to write a paper or a book on, or perhaps it was too complex a subject. I like to congratulate my old friends Bruce and Arun and their co-author for writing such a comprehensive book for all interested parties: shipowners, shipyards, surveyors, and students.

Owing to the relatively high costs in time and money for ship repairs and drydocking, classification societies allow continuous surveys and maintenance such that ships need not go out of service to go to a shipyard unless something below the waterline needs to be repaired. Also, with comprehensive data and record keeping, owners and shipyards can prepare much better in advance of a ship arriving at the shipyard. Surveyors, too, can have at his fingertips all the information about the ship that he needs at the shipyard. Again, this will reduce costs and time out of service.

I wish all a better understanding of the process and to continue finding ways to make it more effective and efficient to maintain the more complex and expensive assets of a modern ship.

Singapore
October 2021

Prof. Choo Chiau Beng
Rector of NUS Residential College 4
National University of Singapore
Former CEO of Keppel Corporation
Former CEO of Keppel Fels, and CEO and Chairman of
Keppel Offshore and Marine

Foreword by Peter G. Noble

For as long as humans have taken to the sea, ships and other crafts have served as a means of material and technological exchange and have embodied the culture that has produced them. The materials to build a ship reflect the ability to procure resources; the construction process demonstrates the ability to organize labour, and the construction techniques define the level of skill and technology available.

The earliest known physical example of a large vessel is the Khufu ship from Egypt. This ship had lain undisturbed since it was sealed in a pit carved into the Giza bedrock approximately 4,500 years before. Continuing archaeological exploration has discovered other ships from these ancient times that show aspects of ship repair and material salvage and rebuilding. However, we can say that ship repairing and conversion business are over 4000 years old with some certainty. At the same time, there are many texts on ship design and shipbuilding, and there has been a dearth of published information on ship repair.

Modern shipbuilding produces some of the most complex assets that we design, construct, and use, and these assets are becoming longer lived as we figure out how to design, build, maintain, and repair better. A 25-year ship life may be the design norm, but in practice, many vessels reach 30, 40, and even 50+ years in service if well built and well maintained. In addition, modern processes and techniques, including asset life cycle management processes, laser scanning systems, robotic inspection services, and improved coating systems, contribute to ships having increased useful working lives.

In 2021, BIMCO, the world's leading international shipping association, estimated the world commercial fleet to stand at approximately 74,500 vessels and projects that by 2025 the number will rise close to 80,000. All these ships will require maintenance and repair on an ongoing basis with regular visits to ship repair yards worldwide.

The authors of this book have done an admirable job in bringing together ship repair data in an organized way. Moreover, they have presented these data in a fashion which will allow the less experienced ship repairer to understand the scope, labour requirements, and scheduling for routine hull maintenance and repairs. While the data used by the authors come from a single Asian ship repair yard, they are helpful, both in providing numerical values for specific sub-topics with the ship repair process and further as a demonstration of how to organize and use such data.

In addition to explaining the data presented, the book's concluding chapters describe the essential topics of ship repairing data collection, management information systems, and human resource, which provide the reader with guidance that will allow for improved data collection and ship repair management.

December 2021   Peter G. Noble
C. Eng, C. MarEng, FIMarEST, FSNAME, FCAE, FIES
Naval Architect, Marine and Ocean Engineer
Houston, TX, USA

Preface

While working in shipyards involving ship repairing, the first two authors have encountered problems regarding confirmation of the work scope to be carried out and the durations (repairing time and drydocking time) of ships. Confirmation issues of work scope are mainly related to structural steel renewal quantity and locations, hull, tank coating (blasting and painting), renewal quantity, piping system renewal length, etc. There were no appropriate answers from the shipowners' representatives and the ships' staff until proper representatives inspected individual items. Problems of duration are related to the time (number of days) allowed by the shipyard to the respective production team for repairing and drydocking. Contractually, this is the mutually agreed time between the owner and shipyard. Production department, commercial department, or ship repairing department (as appropriate for different shipyards) makes the final decision and signs the agreement with the respective owner for the individual ship.

The background of the first problem is understandable. It is impossible to know the appropriate scope of those items mentioned above while the vessel is in service. The only opportunity to see the extent of works is when the ship is in the drydock and inspected. As a result, the actual scope of works always differs from that mentioned in the shipowner's repair specification before a ship's arrival.

Regarding the second problem, previous references of the same ship or similar ship and their scope of works are followed to ascertain the duration of repairing time and drydocking time. Interestingly, the prevailing situations during the specified time in the shipyard are not considered. In real life, though scopes of work govern the repairing duration (number of days) in the shipyard and the drydocking period in the drydock, the shipyard's prevailing situation plays a vital role in the final duration in the shipyard and the drydock. The following examples will describe the problems, and it happens very often.

Case 1: The ship was moored along the quayside before entering the drydock. The crew members and shipyard started preparatory works (maybe one or two days).

Case 2: The ship was placed in the drydock directly on arrival due to preoccupying all quays. The crew member and shipyard did not have any time for preparatory works before entering the drydock.

In case 1, there were times to carry out preparatory works for tail shaft withdrawal. In case 2, there is no time for any preparatory works. So, the preparatory works started in the drydock, and the time required for preparatory works will be a part of the drydocking time. Therefore, the drydocking time in case 2 will be longer than in case 1, considering the precise scope of work and work progress rate.

Case 3: The ship was moored alongside the quay before entering the drydock, and the shipyard started the preparatory works, say, unloaded the hatch covers to shore for access/repair works.

Case 4: The ship was moored alongside another ship (double banking) where the shore crane's access is limited. Therefore, the preparatory works had started only after entering the drydock.

In case 3, there were time and facilities to carry out preparatory works (unloading hatch covers to shore for repairing/access) before entering the drydock. However, in case 4, hatch covers could not be offloaded to shore due to the ship's location. So, the preparatory works had started in the drydock, and the time required for preparatory works will be a part of the drydocking time. Therefore, the drydocking time in case 4 will be longer than in case 3, considering the exact scope of work and work progress rate.

The repairing time in case 1 and case 2 may be the same, but the invoice will be different due to different drydocking times. No doubt, case 2 has the privilege to have a higher invoice. However, many more adverse situations arose, which ultimately affected the repairing and drydocking days. Therefore, not only the ship in question but the ships due for subsequent drydocking are also involved.

Both parties, shipowners and shipyards, expect and try to make shorter repairing and drydocking times in real life. For shipowners, less repairing time means less loss of earning, and less drydocking time means saving cost because drydocking charges are much higher than quayside charges, for shipyards to maximize the number of ships as a whole and in the drydock in a particular period (say yearly), increasing the revenue.

Keeping all the above points in mind, the two authors have attempted to frame a guideline for estimating repairing time, drydocking time, repairing labour, drydocking labour, hull coating renewal areas, and others against various combinations of variables. The authors hope this may help the shipowners, ship managers, and shipyards estimate the required variables.

For any ship repair yard seeking to improve the performance, especially the cost and timescale of the repair work, there is a need to analyse previous contracts and complete information on the shipping market, specific shipowners, and the capabilities within the shipyard and its usual subcontractors. To support the need, the shipyard first requires a comprehensive management information system.

An information system must integrate all aspects of the shipyard operations, making information available at all stages of work, from marketing to preparing and submitting a final invoice. Unfortunately, few such systems exist, except as proprietary information within a single shipyard. The book describes the requirements that a suitable system must satisfy, whether purchased from a supplier or developed in-house.

To make a system effective, there must be reasonable and accurate data collection. This is a more challenging problem than may be anticipated. Data that is not collected immediately after the event which produces the data has occurred. If there is a delay, then the data will degrade. It can be mislaid, forgotten, misrecorded, or otherwise compromised. Poor data is dangerous and can mislead the management to make poor decisions. Accurate production data is particularly important to be able to create accurate costs for invoicing.

Another critical requirement is a well-trained and organized workforce, usually supported by casual work and subcontractors. Typical development of the workforce for a shipyard in the context of the wider industrial environment is described. Recruitment, training, safety, and succession planning are all outlined to offer guidance to shipyard management.

Singapore   Arun Kr Dev
November 2021   Makaraksha Saha
                                                                                                  George Bruce

Contents

Symbols and Abbreviations

$a, b, b_0, b_1, b_2, \ldots$	Regression coefficients
Age	Age of a ship at the time of repair (year)
AP	After perpendicular
AS	Annual survey
A_{BT}	Boot top total area
A_{FB}	Flat bottom total area
A_{HT}	Hull total area
A_{TS}	Topside total area
A_{VB}	Vertical bottom total area
A_{VS}	Vertical side total area ($= A_{BT} + A_{VB}$)
BC	Bulk carrier
BL	Baseline
B_{mld}	Breadth moulded
CaC	Car carrier
CC	Container carrier
ChT	Chemical and product tanker
CL	Longitudinal centre line
CN	Cubic number
COT	Crude oil tanker
COW	Crude oil washing
CS	Classification society
Deadweight	Deadweight of a ship (tonne)
DS	Docking survey
D_{LABOUR}	Drydocking labour (man-day)
D_{mld}	Depth moulded (m)
D_{TIME}	Drydocking time (day)
FP	Forward perpendicular
GC	General cargo carrier
GD	Guarantee docking
GT	Gross tonnage
ICCP	Impressed current cathodic protection

i.w.o.	in the way of
IS	Intermediate survey
k	Number of independent variables used in the respective regression equation
LNG	Liquified natural gas
LNGC	Liquified natural gas carrier
LPG	Liquified petroleum gas
LPGC	Liquified petroleum gas carrier
L_{OA}	Length overall (m)
$L_{OA} * 2(D_{mld} - T_{max})$	Topside area (m^2)
$L_{OA} * 2(T_{max} - T_{max})$	Boot top area (m^2)
$L_{OA} * 2T_{min}$	Vertical bottom area (m^2)
$L_{OA} * B_{mld}$	Flat bottom area (m^2)
$L_{OA} * 2T_{max}$	Vertical side area (m^2)
$L_{OA} * (B_{mld} + 2T_{max})$	Hull total area (m^2)
m	Slope for linear equation
m	Linear length, metre
MD	Main deck
MGPS	Marine growth protection system
n	Sample size
P	Port
Q_{LABOUR}	Quayside labour (man-day)
Q_{TIME}	Quayside time (day)
r^2	Correlation coefficient (for simple regression)
R^2	Coefficient of determination (for multiple regression)
R_{BTB}	Boot top blasting renewal area (m^2)
R_{BTP}	Boot top painting renewal area (m^2)
R_{FBB}	Flat bottom blasting renewal area (m^2)
R_{FBP}	Flat bottom painting renewal area (m^2)
R_{hb}	Hull blasting renewal area (m^2)
R_{hc}	Hull coating renewal area ($= R_{hb} + R_{hp}$) (m^2)
R_{hp}	Hull painting renewal area (m^2)
R_{LABOUR}	Repairing labour (man-day)
R_{LG}	Longitudinal structural members' renewal weight (kg)
R_{MS}	Miscellaneous structural members' renewal weight (kg)
R_p	Pipe renewal length (m)
R_{PG}	Hull plates renewal weight (kg)
R_s	Total structural steel renewal weight (kg)
R_{tb}	Tank blasting renewal area (m^2)
R_{tp}	Tank painting renewal area (m^2)
R_{TG}	Transverse structural members' renewal weight (kg)
R_{TIME}	Repairing time (day)
R_{TSB}	Topside blasting renewal area (m^2)
R_{TSP}	Topside painting renewal area (m^2)
R_{VBB}	Vertical bottom blasting renewal area (m^2)

R_{VBP}	Vertical bottom painting renewal area (m^2)
R_{VSB}	Vertical side blasting renewal area ($= R_{\mathrm{BTB}} + R_{\mathrm{VBB}}$) (m^2)
R_{VSP}	Vertical side painting renewal area ($= R_{\mathrm{BTP}} + R_{\mathrm{VBP}}$) (m^2)
s	Standard deviation
S	Starboard
SS	Special survey
SS	Side shell plates
S_{A}	Age of a ship (year)
S_{D}	Deadweight of a ship (tonne)
S_{T}	Type of a ship (in terms of equivalent numerical value)
Type	Type of a ship
$T_{\max}$	Draft at maximum load line (m)
$T_{\min}$	Draft at light load line (m)

Chapter 1
General Introduction

1.1 Introduction

A ship is a most complex engineering asset. Probably, two significant issues, (i) fulfilling shipowners' requirements and (ii) complying with the operational and environmental (classification society, statutory bodies, flag administration) requirements which are conflicting in nature, make a ship, as a whole, a complex asset. It is more exciting and challenging when a ship is involved in routine maintenance and repair. This challenge is due to the type, age, and deadweight of ships. Different types demand entirely new requirements. In addition, classification society's and flag administration's requirements vary with the age of a ship. So, every repair of a ship is a new project and need to handle accordingly.

1.2 Ship Repairing and Drydocking

The usual meaning of ship repairing is to repair a ship, elaborate components comprising structural steel, hull coating, piping, tank coating, various machinery, equipment, etc. However, in this book, the focus of ship repairing is shifted from execution (to repair items) to quantification (to estimate the scope of repairing items in an appropriate quantity) of repairing scope, duration, workforce (man-days), which will help the shipowners and the shipyards someway.

Generally speaking, 'Docking' of a ship refers to the act of putting a ship in a dock to carry out underwater inspection and maintenance works. The docking facility can be a drydock (graving type), floating dock, slipway, synchro lift, or other facilities. In this book, 'Drydocking' refers to the same act for the same purpose but in a drydock (graving dock) only because data are collected from a shipyard with drydocks only, no other means of docking.

© The Author(s), under exclusive license to Springer Nature Singapore Pte Ltd. 2022
A. K. Dev et al., *Ship Repairing*, Springer Series on Naval Architecture,
Marine Engineering, Shipbuilding and Shipping 12,
https://doi.org/10.1007/978-981-16-9468-4_1

The authors have attempted to introduce a new and practical approach to estimate various ship repairing variables such as repairing time and labour, drydocking time and labour, hull coating renewal area using actual data from the practical field (shipyard).

1.3 Background Knowledge of Users of This Book

In this book, no attempt is made to introduce any new scientific and engineering theory. The whole concept is to use the real-life data collected from the practical field (shipyard), analyse it, and make some conclusions. It is expected that the practical experience in a shipyard involving ship repairing will be enough to understand and use the book.

Types of ships and their technical and operational requirements make the ship repairing subject more interesting. Therefore, reading this book should not be difficult for anyone interested in knowing how different variables like repairing time, repairing labour, structural steel renewal weight, hull coating renewal quantity, etc., respond with age, deadweight, and type.

1.4 Experience Sharing

Both authors have gained shipbuilding and ship repairing experiences through decades of working in shipyards. While working in shipyards, they encountered many similar natures in job quantification and decision making from shipowners' representatives. The situation was almost identical for each ship. Issues are mainly related to the scope of structural steel renewal weight and locations, hull coating renewal area, piping renewal length, tank coating renewal area, etc.

Through this book, the authors intend to share their experiences, thoughts, and useful data to overcome those problems by providing some guidelines for shipowners and shipyards to quantify items that can only be known in a drydock.

1.5 General Layout of the Book

The remaining Chapters are arranged as follows.

Chapter 2 covers data collections and their brief illustrations. Collected data of various variables are presented in tabular forms. These tables will tell how a particular variable is collected against what set of other variables. For example, hull coating renewal work and structural steel renewal work data are collected against age, deadweight, and type. In contrast, repairing time and labour data are collected against the repairing activities in addition to age, deadweight, and type. Collected data are illustrated in graphical forms (figures) in terms of distribution. These figures will provide an instant idea about the distribution and concentration of variables over the sample size. It will also describe the range of individual variables of the sample.

Chapter 3 covers ship repairing time. Ship repairing time, duration of stay of a ship in a shipyard is a part of a ship's routine maintenance schedule, mainly required by the classification societies and the flag states. The shipowner and shipyard always try to reduce the repairing time to reduce the loss of income (for an owner) and maximise the annual turnover through handling more ships (for a shipyard). A multiple linear regression model was developed and analysed using these primary data. Ship repairing time was then expressed as a function of a ship's age, deadweight, repairing works of mainly hull coating, piping, structural steel, and tank coating.

Chapter 4 covers drydocking time. The duration of a ship's stay in a drydock depends on the scope of routine underwater repairing works to be carried out. More specifically, the repairing works that are affected by outside water. These are hull cleaning, coating (blasting and painting), rudder, propeller, stern tube aft seal, hull anodes, ICCP, sea valves, sea chests, tunnel thruster(s), bottom plugs, underwater structural steel and so on. These works dictate ships' drydocking time (days) and labour (ship drydocking labour will be covered in Chapter 6). A multiple linear regression model was developed and analysed using these primary data. Ship drydocking time was then expressed as a function of a ship's age, deadweight, type, and hull coating.

Chapter 5 covers ship repairing labour. Labour cost is an important and sensitive issue in labour-intensive industries. Ship repairing work is, by nature, labour-intensive and not prone to automation. In regular ship repairing or routine maintenance, labour cost contributes the highest amount in the final invoice. This figure may go up to 70% of the total cost and is directly contributed by labour utilised to repair the ship. Lesser man-days can be translated into the lower final invoice and higher productivity, which can help the shipyard stay in a competitive market. A multiple linear regression model was developed and analysed using these primary data. Ship repairing labour was then expressed as a function of a ship's age, deadweight, type hull blasting, hull painting and structural steel.

Chapter 6 covers drydocking labour. Being a part of ship repairing labour, ship drydocking labour has a significant impact on the final invoice. Lesser drydocking labour can be translated into the lower final invoice, the same as ship repairing labour. A multiple linear regression model was developed and analysed using these primary data. Ship drydocking labour was then expressed as a function of a ship's age, deadweight, type, hull blasting and hull painting.

Chapter 7 covers the hull coating renewal area. Hull coating renewal is a part of the routine maintenance works of a ship. It is carried out only when the ship is in a dock. For a regular maintenance schedule, hull coating repairing jobs dictate a ship's stay in the drydock. Therefore, it is vital to get accurate information about the hull coating repairing scope of works before drydocking, to maintain the dock operation schedule, drydocking duration. The physical size of a ship greatly influences the hull blasting, and painting repairing works irrespective of the ship's design parameters.

An approach for preliminary estimation of blasting and painting before drydocking is proposed based on ships' dimensions, the hull's actual area, and the various hull locations. A multiple linear regression model was developed and analysed using primary data. Hull blasting and hull painting were then expressed separately as a function of a ship's age, deadweight, and type.

Chapter 8 covers structural steel renewal weight. Structural steel renewal is carried out as a part of routine maintenance of a ship during its service life. It results from structural deficiencies beyond the limit set by the classification societies, either regarding scantling or structural deformations or both. Reduction in thickness of steel and distortion is due to many natural phenomena. Structural steel replacement cost is the highest in a ship repairing invoice for a shipowner, and it also consumes the maximum resources from a shipyard. Hence, prior information about the scope of structural steel replacement might help the shipowner with proper budget allocation and schedule to meet the financial and commercial commitments. A multiple linear regression model was developed and analysed using these primary data. Structural steel renewal weight was then expressed as a function of a ship's age, deadweight, and type.

Chapter 9 covers structural steel renewal location. Structural steel renewal is a part of maintenance work and is suggested by the classification societies. Prior information about the location of steel renewal works can help the ship owners prepare the ship before going to the shipyard and planning for the required logistics. By doing so, the shipowners will be able to save cost in terms of less idle time in the shipyard, and the latter can also increase the revenue in terms of minimising mobilisation time. Data of renewal locations of selected structural members were analysed and presented in tables and figures to demonstrate the behaviour of renewal locations concerning a ship's dimension appropriate for respective structural members, age, and length of a ship.

Chapter 10 covers the quality of data collection and the importance of its timely entry into the shipyard's management information system (MIS). Data collection is fundamental for effective operations and should be recorded immediately on task completion and be entered into the system, enabling the concerned department to review, analyse, and take necessary actions if yet to be taken. Manual collection and entry of data by respective supervisors has disadvantages in terms of accuracy and delay access due to their other responsibilities. Dedicated staff for data collection and their entry may help to improve the situation. Data collection can be automated to some extent, and this is an excellent way to reduce the possibility of incorrect information entering the system.

Chapter 11 covers management information systems in ship repair. Management systems in ship repair differ from other industries, including ship conversion. It includes marketing, enquiries, estimates of man-hour and materials, contract, production planning, work schedule. A comprehensive database of various tasks may provide a beneficial basis for management decision making efforts. The information management system is fundamental to the successful management of a ship repair contract. The management of a shipyard must analyse their business carefully, consider all the possible alternatives and costs, calculate the benefits of a system, and choose what appears to be the best solution.

Chapter 12 covers human resources management in a shipyard regarding recruitment and training to develop different trades' skills. Being a large and complex organisation, marine production faces a universal problem of recruiting and retaining a high degree of labour specialisation with the right talent. Depending on the company context, the needs, and responses in maintaining a skilled workforce will vary. The intake of human resources is greatly influenced by the location of the company and the product mix. Because the more complex the ships to be repaired, the more prominent the range of skills and higher skill levels required of the workers will be. At the same time, the shipyard management must duly consider occupational health issues of the workforce, environmental issues, and waste management issues.

Chapter 2
Data History

2.1 Introduction

In this Chapter, data are presented in tables, and their general illustrations are presented in graphs. All these data were collected using various documents from a single shipyard in Asia.

2.2 Data Collection

Data of repairing time and labour (refers to time and labour consumed during a ship's stay in a shipyard), drydocking time and labour (refers to time and labour consumed during a ship's stay in a drydock) and various repairing activities were collected from the shipyard using multiple related documents. These data represent practical and real-life information. Collected data are presented in tabular form without ships' names to avoid identifying ships against a fictitious code number assigned to each ship.

The sample size of the collected data of various variables is different. In other words, each sample of data does not cover all mentioned variables. As such, collected data are presented separately for each variable. Furthermore, each data is compiled against the type, age, and deadweight of ships in addition to the respective variables.

2.2.1 Repairing Time (R_{TIME})

Data of repairing time were collected from the ships' movement schedule of the shipyard. This document records the arrival dates, dock-in, dock-out, and departure

Table 2.1 Collected data of repairing time

S. No	Code No	Type of ships	S_A (year)	S_D (tonne)	R_{TIME} (day)
1	7954	Crude oil tanker	5	285,356	20
2	7957	Container carrier	4	66,520	10
3	7959	Container carrier	4	66,520	9
4	7961	Container carrier	4	66,520	10
5	7989	Bulk carrier	9	151,418	8
6	8008	Crude oil tanker	13	87,768	33
7	8014	Crude oil tanker	3	47,172	10
8	8025	Container carrier	4	61,489	10
9	8035	Bulk carrier	5	150,790	7
10	8046	Bulk carrier	2	69,638	6

from the shipyard. A sample of collected data (ten data points) is presented in Table 2.1, which shows the type, age, deadweight and repairing time against a code number.

2.2.2 Drydocking Time (D_{TIME})

Data of drydocking time were collected from the ships' movement schedule of the shipyard. This is the identical record used for repairing time. A sample of collected data (ten data points) is shown in Table 2.2, which shows the type, age, deadweight and drydocking time against a code number.

Table 2.2 Collected data of drydocking time

S. No	Code No	Type of ships	S_A (year)	S_D (tonne)	D_{TIME} (day)
1	8085	Chemical tanker	20	37,067	9
2	8102	Container carrier	8	22,735	16
3	8127	Container carrier	20	38,485	14
4	8145	Container carrier	8	58,986	8
5	8162	Container carrier	3	63,693	5
6	8172	Container carrier	5	59,283	10
7	8186	Crude oil tanker	12	94,009	9
8	8189	L.P.G. carriers	19	30,466	6
9	8205	Bulk carrier	17	39,600	7
10	8229	Crude oil tanker	7	258,079	5

Table 2.3 Collected data of repairing labour

S. No	Code No	Type of ships	S_A (year)	S_D (tonne)	R_{LABOUR} (man-day)
1	9216	Bulk carrier	2	75,707	305
2	9233	Chemical tanker	20	17,740	3226
3	9234	Chemical tanker	10	45,683	2006
4	9237	Crude oil tanker	2	107,113	946
5	9248	LPG carrier	26	49,985	3778
6	9254	Bulk carrier	1	74,759	1283
7	9272	Container carrier	11	24,378	1044
8	9276	Bulk carrier	5	50,326	856
9	9306	Container Carrier	14	59,560	8204
10	9325	Crude Oil Tanker	15	145,200	454

2.2.3   Repairing Labour (R_{LABOUR})

Data of repairing labour were collected from the respective ship's daily workforce (labour/day) record. This document records the number of persons and the duty period (day shift, night shift, or overtime). A sample of collected data (ten data points) is displayed in Table 2.3, showing the type, age, deadweight and repairing labour against the code number.

2.2.4   Drydocking Labour (D_{LABOUR})

Drydocking labour data were collected from the daily workforce record of the respective ship. This is the same record used for repairing labour. A sample of collected data (ten data points) is depicted in Table 2.4, showing the type, age, deadweight and drydocking labour against the code number.

2.2.5   Hull Coating Renewal Area (R_{hb} and R_{hp})

Hull blasting and painting renewal area data were collected from the work completion report of the respective ship. This document records all works carried out onboard the ship in detail. A sample of collected data (ten data points) is demonstrated in Table 2.5, showing the type, age, deadweight, hull total area, hull blasting and painting renewal area against the code number.

Table 2.4 Collected drydocking labour

S. No	Code No	Type of ships	S_A (year)	S_D (tonne)	D_{LABOUR} (man-day)
1	9353	Crude oil tanker	3	299,997	1037
2	9362	Crude oil tanker	8	279,999	1232
3	9363	Container carrier	10	15,421	1609
4	9367	Crude oil tanker	3	106,516	1138
5	9376	Container carrier	20	28,348	3835
6	9396	Bulk carrier	5	53,553	737
7	9410	Chemical tanker	3	33,916	625
8	9412	LPG carrier	14	49,999	2625
9	9466	Container carrier	19	51,534	6206
10	9502	Bulk carrier	13	69,183	2934

Table 2.5 Collected data of hull blasting and painting renewal area

S. No	Code No	Type of ships	S_A (year)	S_D (tonne)	Hull area (m^2)		
					A_{HT}	R_{hb}	R_{hp}
1	9237	Crude oil tanker	2	107,113	18,350	675	25,144
2	9306	Container carrier	14	59,560	18,628	14,157	70,561
3	9352	Crude oil tanker	15	258,080	34,400	964	56,588
4	9353	Crude oil tanker	3	299,997	37,080	300	57,752
5	9362	Crude oil tanker	8	279,999	32,980	1439	43,287
6	9363	Container carrier	10	15,421	6840	976	12,020
7	9367	Crude oil tanker	3	106,516	18,727	805	27,204
8	9376	Container carrier	20	28,348	8900	5275	24,036
9	9380	Crude oil tanker	10	126,646	23,450	1717	69,966
10	9396	Bulk carrier	5	53,553	11,460	2364	38,086

2.2.6 Structural Steel Renewal Weight (R_S)

Structural steel renewal weight data were collected from the work completion report, structural steel renewal plan (drawings) and steel material fabrication list recorded by the mould-loft shop. These documents, especially the steel material fabrication list, record the structural members' size and weight. A sample of collected data (ten data points) is highlighted in Table 2.6, showing the type, age, deadweight, and structural steel renewal weight against the code number.

Table 2.6 Collected data of structural steel renewal weight

S. No	Code No	Type of ships	S_A (year)	S_D (tonne)	R_S (kg)
1	9237	Crude oil tanker	2	107,113	544
2	9325	Crude oil Tanker	15	145,200	5174
3	9353	Crude oil tanker	3	299,997	184
4	9362	Crude oil tanker	8	279,999	1869
5	9363	Container carrier	10	15,421	373
6	9367	Crude oil tanker	3	106,516	482
7	9376	Container carrier	20	28,348	18,215
8	9380	Crude oil tanker	10	126,646	3188
9	9396	Bulk carrier	5	53,553	377
10	9410	Chemical tanker	3	33,916	537

Table 2.7 Collected data of piping renewal length

S. No	Code No	Type of ships	S_A (year)	S_D (tonne)	R_P (m)
1	9237	Crude oil tanker	2	107,113	3.0
2	9380	Crude oil tanker	10	126,646	323.0
3	9402	Crude oil tanker	5	107,132	20.0
4	9419	Crude oil tanker	9	106,681	18.2
5	9424	Crude oil tanker	14	156,835	69.7
6	9431	Crude oil tanker	15	98,640	368.5
7	9458	Crude oil tanker	5	99,999	1.4
8	9463	Crude oil tanker	13	265,353	185.0
9	9490	Crude oil tanker	5	104,875	26.5
10	9498	Crude oil tanker	20	151,803	27.7

2.2.7 Piping Renewal Length (R_P)

Data of piping renewal length were collected from the respective ship's work completion reports. A sample of collected data (ten data points) is introduced in Table 2.7, which shows the type, age, deadweight, and piping renewal length against the code number. Piping renewal length is counted as running meter irrespective of bends, diameter, wall thickness (schedule) and the system.

2.2.8 Tank Coating Renewal Area (R_{tb} and R_{tp})

Tank blasting and painting renewal area data were collected from the respective ship's work completion report. A sample of collected data (ten data points) is produced in

Table 2.8 Collected data of tank blasting and painting renewal area

S. No	Code No	Type of ships	S_A (year)	S_D (tonne)	R_{tb} (m^2)	R_{tp} (m^2)
1	9380	Crude oil tanker	10	126,646	2520	9380
2	9431	Crude oil tanker	15	98,640	24,616	54,155
3	9518	Crude oil tanker	10	156,835	2602	6245
4	9527	Crude oil tanker	17	69,996	5300	51,304
5	9540	Crude oil tanker	9	308,491	1176	2717
6	9574	Crude oil tanker	10	309,996	2400	5760
7	9608	Crude oil tanker	10	309,996	4404	28,369
8	9649	Crude oil tanker	10	310,138	2468	11,844
9	9684	Crude oil tanker	15	147,564	3300	8280
10	9701	Crude oil tanker	10	158,000	12,623	38,897

Table 2.8, showing the type, age, deadweight, tank blasting, and painting renewal area against the code number. Tank coating renewal works are collected irrespective of tanks (cargo tank or ballast tank).

2.2.9   *Structural Steel Renewal Locations*

Structural steel renewal location data were identified using the ship's structural steel renewal plan and ship drawings. A sample of collected data (ten data points) is exhibited in Table 2.9, showing the type, age, deadweight, and the number of individual structural members' renewal locations against the code number. Location is defined as in between two consecutive frames and shell longitudinals for side shell plate (for example), and if exceeded the boundary, considered additional location.

2.3   Illustrations

In this section, data are illustrated in graphs for different independent and dependent variables (age, deadweight, type, repairing time and labour, drydocking time and labour) and repairing activities (hull blasting and painting, structural steel renewal weight and location), in term of distribution. From these illustrations, the reader will understand how different variables are distributed over the sample data. This will also indicate the concentration of variables.

Table 2.9 Collected data of structural steel renewal locations

Part-I

Col-A	Col-B	Col-C	Col-D	Col-E	Col-F	Col-G	Col-H	Col-I
S. No	Code No	Type of ships	S_A (year)	S_D (tonne)	Deck plate	Side shell plate	Bottom plate	Tank top plate
1	9237	Crude oil tanker	2	107,113	0	1	0	0
2	9325	Crude oil tanker	15	145,200	1	1	0	0
3	9353	Crude oil tanker	3	299,997	0	0	0	0
4	9362	Crude oil tanker	8	279,999	0	0	0	0
5	9363	Container carrier	10	15,421	0	0	1	0
6	9367	Crude oil tanker	3	106,516	0	1	0	0
7	9376	Container carrier	20	28,348	0	3	0	17
8	9380	Crude oil tanker	10	126,646	1	1	0	0
9	9396	Bulk carrier	5	53,553	0	0	0	0
10	9410	Chemical tanker	3	33,916	0	0	0	0

Part-II

Col-J	Col-K	Col-L	Col-M	Col-N
Topside tank bottom plate	Deck longitudinal	Side shell longitudinal	Bottom longitudinal	Double bottom tank under deck longitudinal
0	0	1	0	0
0	1	2	0	0
0	0	0	1	0
0	0	7	0	0
0	0	0	1	0
0	0	0	0	0
0	0	7	0	0
0	0	0	0	0

(continued)

Table 2.9 (continued)

Part-II

Col-J	Col-K	Col-L	Col-M	Col-N
Topside tank bottom plate	Deck longitudinal	Side shell longitudinal	Bottom longitudinal	Double bottom tank under deck longitudinal
0	0	0	0	0
0	0	0	0	0

Part-III

Col-O	Col-P	Col-Q	Col-R	Col-S
Topside tank bottom longitudinal	Longitudinal bulkhead	Longitudinal bulkhead longitudinal	Deck transverse	Bottom transverse
0	0	0	0	0
0	0	0	0	0
0	0	0	0	0
0	0	0	0	0
0	0	0	0	0
0	0	0	0	0
0	12	5	0	0
0	0	0	0	0
0	0	0	0	0
0	0	0	0	0

Part-IV

Col-T	Col-U	Col-V	Col-W
Transverse bulkhead	Side frame i.w.o side shell plate	Side frame i.w.o longitudinal bulkhead	Stringer plate
0	0	0	0
0	1	0	0
0	0	0	0
0	0	0	0
0	0	0	0
0	0	0	0
37	16	0	7
0	0	0	0
0	0	0	0
0	0	0	0

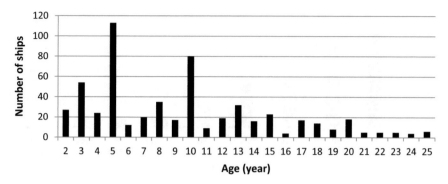

Fig. 2.1 Distribution of age

2.3.1 Age (S_A)

The age of a ship is the time (no. of years) for which she has been in service since delivery. Figure 2.1 demonstrates the distribution of age of sample ships. The average age of the collected sample is ten (10) years, irrespective of deadweight and type.

2.3.2 Deadweight (S_D)

In standard practice, the deadweight defines a ship's size, including the cargo-carrying capacity. Figure 2.2 presents the distribution of deadweight of sample ships. The average deadweight of the collected sample is around 77,500 tonnes, irrespective of age and type.

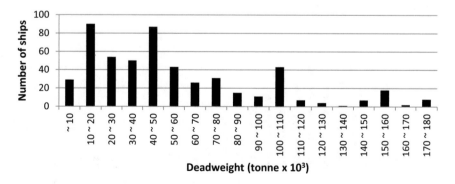

Fig. 2.2 Distribution of deadweight

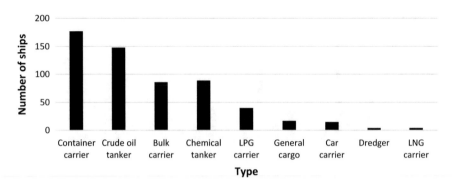

Fig. 2.3 Distribution of type

2.3.3   *Type (S$_T$)*

The type of a ship mainly refers to the type or nature of cargo it carries. For example, the usual merchant ships are crude oil tankers, container carriers, bulk carriers, chemical tankers, product tankers, LPG carriers, LNG carriers, dredgers, car carriers, general cargo carriers, etc. Figure 2.3 validates the distribution of the type of sample ships.

2.3.4   *Repairing Time (R$_{TIME}$)*

A ship's repairing time refers to the duration of stay in a shipyard for routine maintenance works. Figure 2.4 highlights the distribution of repairing time of sample ships. The average repairing time of the collected sample is about 16.0 days, irrespective of age, deadweight, type and repairing activities.

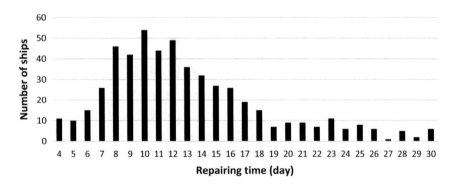

Fig. 2.4 Distribution of repairing time

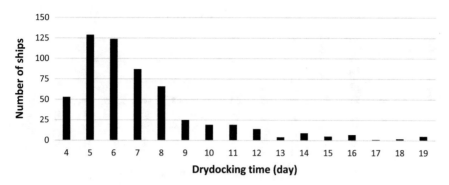

Fig. 2.5 Distribution of drydocking time

2.3.5 *Drydocking Time (D*TIME*)*

Drydocking time of a ship refers to the number of days of stay in a drydock. Figure 2.5 displays the distribution of drydocking time of sample ships. The average drydocking time of the collected samples is 7.25 days irrespective of age, deadweight, type and drydocking activities.

2.3.6 *Repairing Labour (R*LABOUR*)*

The ship repairing labour refers to the workforce (man-day) utilised to complete the repairing activities during the repairing time in the shipyard. Figure 2.6 depicts the distribution of repairing labour of sample ships. The average repairing labour of the collected sample is around 2,750 man-days irrespective of age, deadweight, type, repairing time and repairing activities.

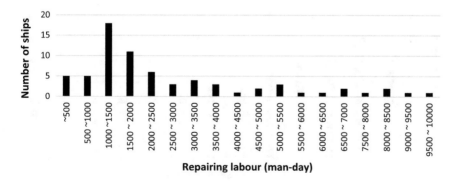

Fig. 2.6 Distribution of repairing labour

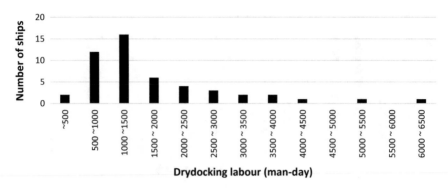

Fig. 2.7 Distribution of drydocking labour

2.3.7 Drydocking Labour (D_{LABOUR})

The drydocking labour refers to the workforce (man-day) utilised during the drydocking period. Figure 2.7 introduces the distribution of drydocking labour of sample ships. The average drydocking labour of the collected samples is about 1750 man-days irrespective of age, deadweight, type and drydocking activities.

2.3.8 Hull Blasting Renewal Area (R_{hb})

Hull blasting of a ship is a surface preparation process, and its renewal work is a part of the hull coating renewal work. Figure 2.8 determines the distribution of the hull blasting renewal area of sample ships. The collected sample's average hull blasting renewal area is around 3000 m^2 (14% of hull total area) irrespective of age, deadweight, and type.

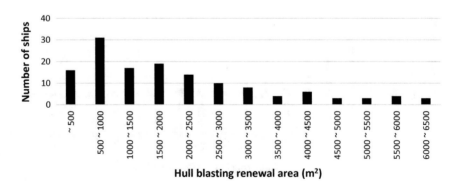

Fig. 2.8 Distribution of hull blasting renewal area

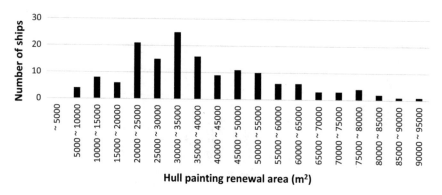

Fig. 2.9 Distribution of hull painting renewal area

2.3.9 Hull Painting Renewal Area (R_{hp})

The hull painting renewal work is a part of hull coating renewal work. After hull blasting, painting work is carried out. Figure 2.9 proves the distribution of hull painting renewal area of sample ships. The collected sample's average hull painting renewal area is about 39,000 m² (187% of hull total area) irrespective of age, deadweight, and type.

2.3.10 Structural Steel Renewal Weight (R_S)

Generally, any structural steel replacement/renewal of a ship is recommended or decided by the attending surveyor of the classification society and agreed upon by the owner, after a close-up inspection of tanks/holds, external hull (side and bottom) in a drydock, depending on the ship's age. Figure 2.10 reveals the distribution of structural

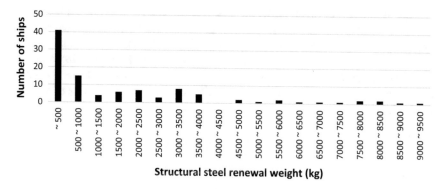

Fig. 2.10 Distribution of structural steel renewal weight

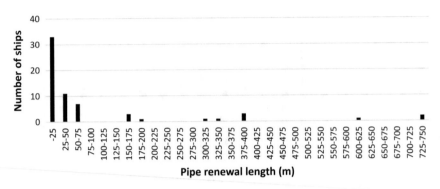

Fig. 2.11 Distribution of pipe renewal length

steel renewal weight of sample ships. The collected sample's average structural steel renewal weight is about 4600 kg irrespective of age, deadweight, and type.

2.3.11 Pipe Renewal Length (R_P)

Generally, a ship's pipe replacement/renewal is recommended or decided by the shipowner's representative except for the pipes under the classification society (pipes penetrating the hull plate like overboard discharge pipes) after a close-up inspection or testing. Figure 2.11 exhibits the distribution of pipe renewal length of sample ships. The collected sample's average pipe renewal length is 89.0 m, irrespective of age, deadweight, and type.

2.3.12 Tank Blasting Renewal Area (R_{tb})

Though it is not a drydocking work, tank blasting (cargo tank and ballast tank), like hull blasting, is a surface preparation process carried out at a regular interval based on coating life or earlier on physical inspection. Figure 2.12 exhibits the distribution of the tank blasting renewal area of sample ships. It is important to note that all samples belong to the crude oil tanker group. The collected sample's average tank blasting renewal area is approximately 6199 m^2 regardless of age and deadweight.

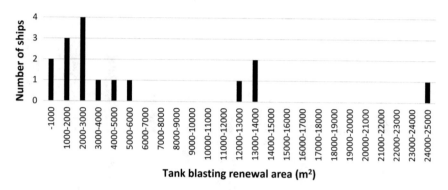

Fig. 2.12 Distribution of tank blasting renewal area

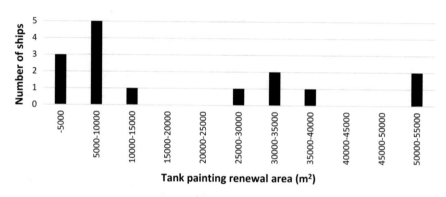

Fig. 2.13 Distribution of tank painting renewal area

2.3.13 Tank Painting Renewal Area (R_{tp})

The tank painting renewal work is carried out after blasting works, the same as hull painting. But painting scheme varies from one coat system to three coat system based on cargo to be loaded in respective tanks. Figure 2.13 displays the distribution of the tank painting renewal area of sample ships. The collected sample's average tank painting renewal area is about 20,063 m^2 regardless of age and deadweight.

2.3.14 Structural Member Renewal Location

The renewal locations of structural steel members were determined concerning the ship's centreline (longitudinal), midship transverse line and baseline for each ship. Figures 2.14 and 2.15 demonstrate the distribution of structural members (by several ships and several renewal locations), respectively, of sample ships.

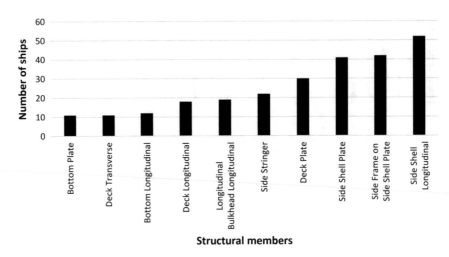

Fig. 2.14 Distribution of structural member renewal by the number of ships

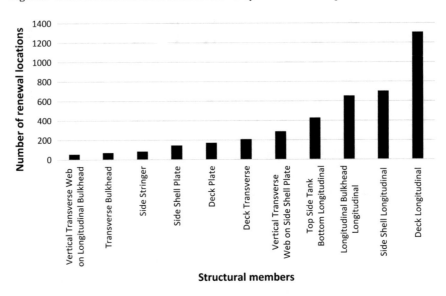

Fig. 2.15 Distribution of structural member renewal locations

Chapter 3
Ship Repairing Time

3.1 Introduction

A ship's repairing time may be defined as the duration of a ship's stay in a shipyard for routine maintenance works. It is counted from the ship's arrival date to the shipyard to departure date from the shipyard, including drydocking time. Drydocking time will be dealt with in a separate Chapter. During this time, the routine maintenance works under the hull, machinery and electrical are carried out under the supervision of the attending shipowner's superintendent(s), ship's crews, and the attending surveyor of the classification society as appropriate. The rules and regulations of the classification societies, flag states and statutory bodies mainly require the routine maintenance of a ship. Both the flag state and classification society call for a ship in a shipyard for different inspections under various survey categories. Surveys are classified as annual surveys, docking surveys, intermediate surveys, and special surveys. Their interval varies from 1 to 5 years. The type of survey applicable to a ship is decided by the classification society based on the ship's age at the time of repairing/drydocking. The ship owners must fulfil the requirements and comply with the rules and regulations for maintaining the ship's statutory certification.

Typical maintenance works are almost like different ships, but their work scope varies with the kind of survey. Tables 3.1 and 3.2 summarise the main maintenance activities of different types of ships. These activities are categorised into two, (1) routine maintenance (Table 3.1) and (2) occasional maintenance (Table 3.2). Routine maintenance refers to works that are carried out regularly and as per CS's requirements, such as hull coating, various clearance measurements, sea valves overhauling, anchor chain calibration, chain locker cleaning, etc. The scope of works varies with age stipulated in the CS's rules and regulations. Occasional maintenance refers to works that are also required by CS's rules (as per survey status of the ship in question) or recommended by the attending CS's surveyor (not mandatory in every routine repairing and drydocking), such as propeller removal, tail shaft withdrawal, tunnel thruster(s) overhaul, structural steel replacement and so on.

© The Author(s), under exclusive license to Springer Nature Singapore Pte Ltd. 2022   23
A. K. Dev et al., *Ship Repairing*, Springer Series on Naval Architecture,
Marine Engineering, Shipbuilding and Shipping 12,
https://doi.org/10.1007/978-981-16-9468-4_3

Table 3.1 Major routine maintenance activities

S. No	Type of ships →	COT	CC	BC	ChT	LPGC	LNGC	GC	CaC
	Routine items↓	Applicability of items							
1	Hull coating	√	√	√	√	√	√	√	√
2	Rudder pintle and bush	√	√	√	√	√	√	√	√
3	Rudder leading edge welding	√	√	√	√	√	√	√	√
4	Propeller polishing	√	√	√	√	√	√	√	√
5	Propeller repairing in place	√	√	√	√	√	√	√	√
6	Stern tube seal	√	√	√	√	√	√	√	√
7	Rope guard	√	√	√	√	√	√	√	√
8	Sea valve	√	√	√	√	√	√	√	√
9	Sea chest	√	√	√	√	√	√	√	√
10	MGPS	√	√	√	√	√	√	√	√
11	Hull anodes	√	√	√	√	√	√	√	√
12	Anchor and chain	√	√	√	√	√	√	√	√
13	Chain locker	√	√	√	√	√	√	√	√
14	Bottom plugs	√	√	√	√	√	√	√	√
15	Ram door	X	X	X	X	X	X	X	√

Table 3.2 Major occasional maintenance activities

S. No	Type of Ships →	COT	CC	BC	ChT	LPGC	LNGC	GC	CaC
	Occasional items↓	Applicability of items							
1	Rudder unship	√	√	√	√	√	√	√	√
2	Propeller withdrawal	√	√	√	√	√	√	√	√
3	Tailshaft withdrawal	√	√	√	√	√	√	√	√
4	Structural steel replacement	√	√	√	√	√	√	√	√
5	ICCP anodes	√	√	√	√	√	√	√	√
6	Cargo tank coating renewal	√	X	X	√	X	X	X	X
7	Cargo holds coating renewal	X	√	√	X	X	X	X	X
8	Stern thruster	√	√	√	√	√	√	√	√
9	Bow thruster	√	√	√	√	√	√	√	√
10	Active fin stabiliser	√	√	√	√	√	√	√	√
11	Draught gauge	√	√	√	√	√	√	√	√
12	Echosounder	√	√	√	√	√	√	√	√
13	Speed log	√	√	√	√	√	√	√	√
14	Bottom pitting	√	√	√	√	√	√	√	√
15	Thickness gauging	√	√	√	√	√	√	√	√

Note √ = Applicable; X = Not Applicable

It is well understood that the repairing time is planned based on the estimated scope of works. The scope of works is generally prepared by the ship superintendent(s) in charge of the ship, with the help of ship crews. The shipyard uses this work scope to organise estimations like schedule, quotation, material procurements, equipment mobilisation, workforce (man-days) allocation, etc. But the reality is different in most cases. Before the ship arrives at the shipyard, a meeting occurs with the superintendent to confirm the scope of works. In most cases, a significant change in the scope of works takes place.

As such, the shipyard must adjust the plan due to the change, decrease or increase. It is believed that there are financial and commercial implications (allocating budget and preparing the ship for service) on the superintendent if the work scope rises significantly. However, there is always a change in the scope of works, which decrease or increase. Issues will be addressed from a practical point of view through analysing the repairing time and corresponding variables.

There are many studies about financial implications on operating expenditures, fleet maintenance cost from various viewpoints and global market conditions, notably oil and gas prices. But very little on the ship repairing time. However, some studies were carried out from different viewpoints and using different variables. Apostolidis et al. [1] highlighted the drydocking cost for tankers. This study explores and identifies the relationship between the drydocking cost and other variables responsible for drydocking cost. The authors then propose a mathematical model for the estimation of the drydocking cost. This drydocking cost referred to the routine repairing cost, not the damage repairing cost. Dev and Saha [2] studied ship repairing time (total days counting from the arrival at the yard to the departure from the yard). It shows that the ship repairing time (day) is linearly related to ships' age, deadweight and repairing works, namely, external hull coating renewal area, structural steel renewal weight, tank coating renewal area and piping renewal length. A mathematical model was developed and proposed a multiple linear regression equation to estimate expected ship repairing time for crude oil tankers using age, deadweight, and quantity of repairing works. Dev and Saha [3] investigated ship repairing time and labour jointly for a similar ship. This paper explores and identifies the possible independent variables responsible for ship repairing time (day) and labour (man-days). It also suggests a possible relationship between various variables in the form of a mathematical equation using multiple linear regressions and verified with statistical testing parameters to demonstrate the adequacy of the model for the system. The condition $f > f_\alpha$ suggests the rejection of the null hypothesis. A mathematical model is developed and proposed a multiple linear regression equation to estimate expected ship repairing time and labour using age, deadweight, and quantity of repairing works. Surjandari and Novita [4] examined drydocking duration. It explores and identifies the drydocking works responsible for the drydocking duration and their relationship using the data mining approach, i.e., CART (classification and regression tree). The authors then propose the linear model to estimate the drydocking time using dock works as input. This model is limited to the drydocking time only. Jose [5] analysed drydocking time and cost and used multicriteria decision-making methods called the goal programming model to minimise the drydocking time and cost. This article demonstrates the technique of the goal programming model to balance the time and cost of drydocking of a ship. Emblemsvag [6] demonstrated the Lean Project Planning (LPP) technique in shipbuilding for project planning. The LPP combines the Earned Value Method and the Last Planner System, which deal with planning, execution, monitoring, and correction. This article presents a case study of a platform supply vessel using the LPP technique. In Victoria et al. [7], it was shown insight into the key findings of the project entitled "Customisation of Web-Based Planning and Production Engineering Technologies to Support Integrated Shipyard Work Flow" initiated by the National Shipbuilding Research Program (NSRP) to enable shipyards to achieve a reduction in project costs and cycle time through project standardisation and the ability to perform replanning. This article also quantifies and appreciates the resulting cost–benefits experienced by each participating shipyard.

This Chapter focuses on how the age, deadweight, type, and routine maintenance work scopes of a ship influence the repairing time individually and jointly. Some assumptions will be made and subsequently be verified through analysing the respective variable of sample data. In this analysis, the repairing time is considered the dependent variable, and others thought independent variables. Finally, a mathematical model is presented to predict the expected repairing time against various independent variables like age, deadweight, type and work scopes and graphs to estimate repairing time against multiple independent variables.

The general assumption is that repairing time is a function of age, deadweight, (age * deadweight), type, hull blasting renewal area, hull painting renewal area, hull coating renewal area, structural steel renewal weight, piping renewal length, tank blasting renewal area, tank painting renewal area, tank coating renewal area and they are linearly associated. In other words, age, deadweight, (age * deadweight), type, hull blasting, painting, coating renewal area, structural steel renewal weight, piping renewal length, tank blasting, painting, coating renewal area each influence repairing time having a linear relationship. It is essential to highlight that the mentioned repairing activity, except tank coating, is the most common quantifiable event in every routine maintenance schedule. Each of them will be discussed in the following sections.

3.2 Repairing Time Versus Age (R_{TIME} vs S_A)

At the time of repairing, the age of a ship is the time (no. of years) for which she is in service. It is counted from the ship's date of delivery from its builder to its owner and can be considered the date of birth of a ship. This date is used in all documents/certificates issued for the first time by the underwriter, the flag state, the classification society, etc. As such, an older ship will have a longer service life. It is thus expected that older ships with older machinery and equipment will experience higher wear and tear depending on the shipowner's maintenance policy. Also, the flag state rules and the CS's rules demand higher standards/criteria of testing, inspection, or survey for older ships. In the end, older ships need more extensive repair and

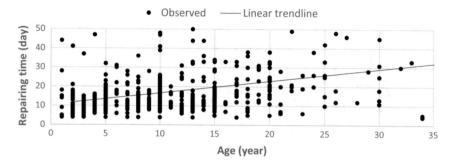

Fig. 3.1 Repairing time versus age

maintenance than newer ones. Therefore, an older ship will require more time than a newer one and is assumed to be linearly related.

Initial investigation of repairing time versus age is demonstrated in Fig. 3.1. It shows the behaviour of repairing time against age and a strong and positive linear relationship. The linear equation, $R_{TIME} = 10.554 + 0.616 * S_A$, provides the best goodness of fit to the sample data with a correlation coefficient of 0.428. Therefore, the assumption that the repairing time is a function of age irrespective of deadweight and type and linearly associated is valid. More specifically, older ships are expected to demand a longer repairing time than newer ones.

3.3 Repairing Time Versus Deadweight (R_{TIME} vs S_D)

A ship's size can be defined by its dimensions (length, breadth, and depth) or capacity (gross tonnage, net tonnage, or deadweight). In usual practice, the deadweight defines a ship's size, including the cargo carrying capacity. In this book, deadweight is used as a size. A big size ship means a higher deadweight means more significant dimensions with larger machinery and equipment. Logically, the bigger ships need longer repairing time for routine maintenance works. Therefore, a bigger ship will require more time than a smaller one and is anticipated to be linearly related.

Initial examination of repairing time versus deadweight is presented in Fig. 3.2. It shows the behaviour of repairing time against the deadweight and a positive linear relationship. The linear equation, $R_{TIME} = 12.614 + 0.024 * (S_D/10^3)$, yields the best goodness of fit to the sample data with a correlation coefficient of 0.215. Therefore, the assumption that the repairing time is a function of deadweight irrespective of age and type and linearly associated is valid. More clearly, bigger ships are likely to demand a longer repairing time than smaller ones.

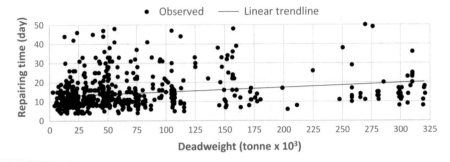

Fig. 3.2 Repairing time versus deadweight

3.3.1 Repairing Time Versus (age * deadweight) [R_TIME vs (S_A*S_D)]

The repairing time is investigated against a new variable, the product of age and deadweight (age * deadweight). The simple logic behind this is that when repairing time is examined against age, the effect of deadweight is ignored. Similarly, when repairing time is investigated against deadweight, the impact of age is disregarded. The use of the product of age and deadweight will allow to include both simultaneously, and the result will be more representative. Therefore, ships with a higher (age * deadweight) value will require longer repairing time irrespective of type, works scope and assumed to be linearly associated.

Initial analysis of repairing time and (age * deadweight) is confirmed in Fig. 3.3 showing the repairing time against (age * deadweight) irrespective of type. The figure shows a positive relationship. This is likely because the higher (age * deadweight) demands bigger size and older ships and eventually, longer repairing time. The linear equations, $R_{TIME} = 5.961 + 8.089 * [(S_A * S_D)/10^6]$, delivers the best goodness of fit relationship with a correlation coefficient of 0.668.

Therefore, the assumption that repairing time is a function of (age * deadweight) irrespective of the type of ship and linearly associated is valid. Particularly, bigger,

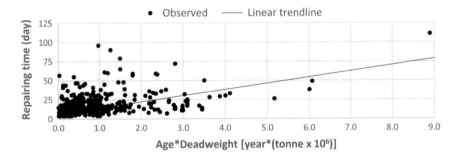

Fig. 3.3 Repairing time versus (age * deadweight)

and older ships are estimated to require longer repairing time than smaller and newer ships.

3.4 Repairing Time Versus Type (R_{TIME} vs S_T)

The type of a ship mainly refers to the type or nature of cargo it carries. The usual types of merchant ships are crude oil tankers, container carriers, bulk carriers, chemical tankers, product tankers, LPG carriers, LNG carriers, dredgers, car carriers, general cargo carriers, etc. Because of the type of cargo and nature of cargo, the configuration of the respective ship varies widely, including machinery and equipment. Also, there are some inherent differences between ships concerning machinery and equipment, piping arrangement, tank arrangement, geometrical configurations, cargo handling facilities, etc. Due to this fact, it is also logical to expect that the different types of ships may require different repairing times. Therefore, diverse ships will require different repairing times, even though they are of similar age and deadweight and likely to be linearly involved.

An initial study of repairing time versus the type of ships is established in Fig. 3.4. It shows a linear relationship. Therefore, the assumption that the repairing time is a function of type irrespective of age and deadweight and linearly associated is valid. More specifically, different ships will require different repairing times even if they are of the same age and deadweight.

3.4.1 Repairing Time Versus Age (R_{TIME} vs S_A) for Types

Data used in Fig. 3.4 were further explored against age for crude oil tankers, container carriers, chemical tankers, bulk carriers, liquified petroleum gas carriers, general

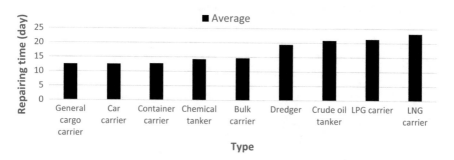

Fig. 3.4 Average repairing time versus type

cargo carriers and car carriers. Results are depicted in Figs. 3.5, 3.6, 3.7, 3.8, 3.9, 3.10 and 3.11. The primary characteristics of all these figures are like that of Fig. 3.1 but with different magnitudes of response.

Table 3.3 summarises the trendline equations and correlation coefficients of repairing time-age relationship under linear form for crude oil tankers, container carriers, chemical tankers, bulk carriers, liquified petroleum gas carriers, general

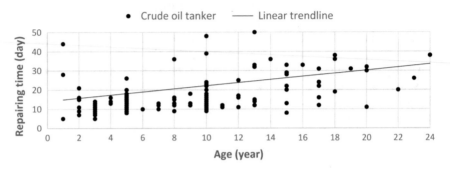

Fig. 3.5 Repairing time versus age for crude oil tankers

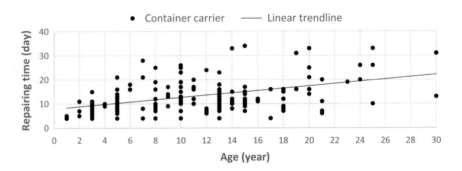

Fig. 3.6 Repairing time versus age for container carriers

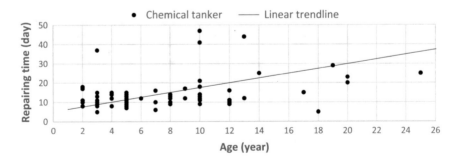

Fig. 3.7 Repairing time versus age for chemical tankers

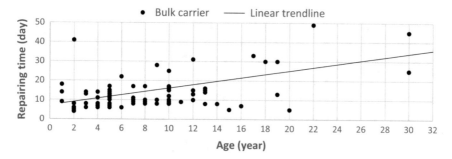

Fig. 3.8 Repairing time versus age for bulk carriers

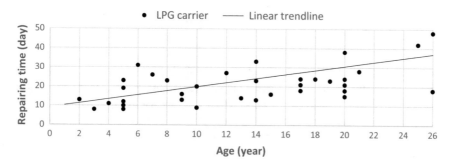

Fig. 3.9 Repairing time versus age for liquified petroleum gas carriers

Fig. 3.10 Repairing time versus age for general cargo carriers

cargo carriers and car carriers. It shows a higher correlation coefficient for types except for crude oil tankers (Fig. 3.5), bulk carriers (Fig. 3.8) and general cargo carriers (Fig. 3.10) when compared with the combined relationship of all types (Fig. 3.1).

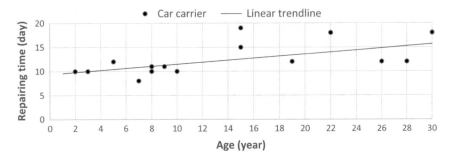

Fig. 3.11   Repairing time versus age for car carriers

3.4.2   Repairing Time Versus Deadweight (R_{TIME} vs S_D) for Types

Data used in Fig. 3.4 were further evaluated against deadweight for crude oil tankers, container carriers, chemical tankers, bulk carriers, liquified petroleum gas carriers, general cargo carriers and car carriers. Results are shown in Figs. 3.12, 3.13, 3.14, 3.15, 3.16, 3.17 and 3.18. The primary behaviour of all these figures is like that of Fig. 3.2 but with different magnitudes of response.

Table 3.4 summarises the trendline equations and correlation coefficients of repairing time-deadweight relationships under linear form for crude oil tankers, container carriers, chemical tankers, bulk carriers, liquified petroleum gas carriers, general cargo carriers and car carriers. It exhibits an improvement in correlation coefficient for types except for crude oil tankers (Fig. 3.12) and general cargo carriers (Fig. 3.17) when compared with the combined relationship of all types (Fig. 3.2).

Table 3.3   Summary of trendline equations and correlation coefficients of repairing time-age relationship under a linear form

Figure No	Trendline equations	r^2	Types
3.5	$Y = 14.089 + 0.809 * X$	0.332	Crude oil tanker
3.6	$Y = 7.852 + 0.480 * X$	0.507	Container carrier
3.7	$Y = 5.135 + 1.239 * X$	0.500	Chemical tanker
3.8	$Y = 7.281 + 0.889 * X$	0.151	Bulk carrier
3.9	$Y = 9.212 + 1.066 * X$	0.462	LPG carrier
3.10	$Y = 9.610 + 0.290 * X$	0.235	General cargo carrier
3.11	$Y = 9.359 + 0.211 * X$	0.407	Car carrier

Note X = Age (year), Y = Repairing time (day)

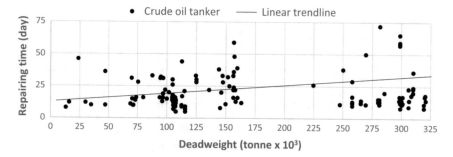

Fig. 3.12 Repairing time versus deadweight for crude oil tankers

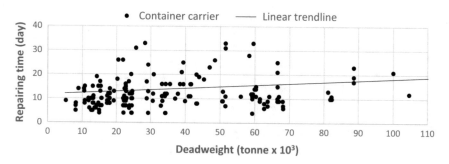

Fig. 3.13 Repairing time versus deadweight for container carriers

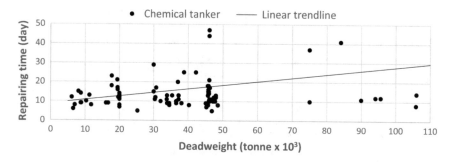

Fig. 3.14 Repairing time versus deadweight for chemical tankers

3.5 Repairing Time Versus Hull Blasting Renewal Area (R_{TIME} vs R_{hb})

Hull blasting renewal work is a part of hull coating renewal work. It is likely that more hull blasting renewal areas will demand more repairing time irrespective of age, deadweight and type and assumed to have a linear relationship.

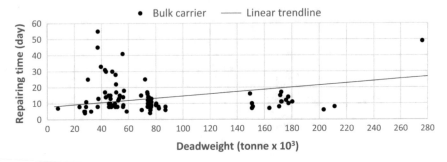

Fig. 3.15 Repairing time versus deadweight for bulk carriers

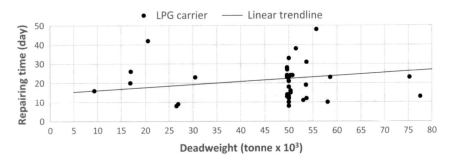

Fig. 3.16 Repairing time versus deadweight for liquified petroleum carriers

Fig. 3.17 Repairing time versus deadweight for general cargo carriers

Initial investigation of repairing time versus hull blasting renewal area is verified as shown in Fig. 3.19, which shows the behaviour of repairing time against hull blasting renewal area. It offers a positive linear relationship. The linear equation, $R_{TIME} = 14.473 + 1.077 * (R_{hb}/10^3)$, forecasts the best goodness of fit to the sample data with a correlation coefficient of 0.412.

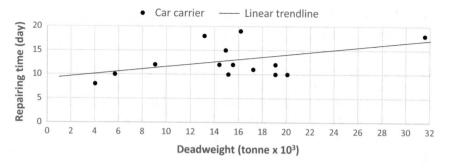

Fig. 3.18 Repairing time versus deadweight for car carriers

Table 3.4 Summary of trendline equations and correlation coefficients of repairing time-deadweight relationship under a linear form

Figure No	Trendline equations	r^2	Types
3.12	$Y = 12.780 + 0.065 * X$	0.134	Crude oil tanker
3.13	$Y = 11.697 + 0.065 * X$	0.499	Container carrier
3.14	$Y = 8.903 + 0.188 * X$	0.889	Chemical tanker
3.15	$Y = 7.916 + 0.067 * X$	0.211	Bulk carrier
3.16	$Y = 14.592 + 0.155 * X$	0.230	LPG carrier
3.17	$Y = 9.787 + 0.301 * X$	0.042	General cargo carrier
3.18	$Y = 9.127 + 0.247 * X$	0.245	Car carrier

Note X = Deadweight (tonne/10^3), Y = Repairing time (day)

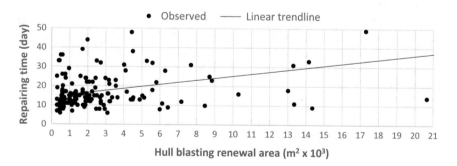

Fig. 3.19 Repairing time versus hull blasting renewal area

Therefore, the assumption that the repairing time is a function of the hull blasting renewal area and linearly associated is valid. So, a higher hull blasting renewal area will demand longer repainting time.

3.6   Repairing Time Versus Hull Painting Renewal Area (R_{TIME} vs R_{hp})

Like hull blasting, hull painting renewal work is also a part of hull coating renewal work, and more hull painting renewal works will require more time. Therefore, more hull painting renewal areas will demand more repairing time irrespective of age, deadweight and type and are anticipated to be linearly associated.

The inspection of repairing time versus hull painting renewal area is validated in Fig. 3.20, which shows the repairing time against the hull painting renewal area. It offers a positive linear relationship. The linear equation, $R_{\text{TIME}} = 11.814 + 0.144 * (R_{\text{hp}}/10^3)$, predicts the best goodness of fit to the sample data with a correlation coefficient of 0.576.

Therefore, the assumption that the repairing time is a function of the hull painting renewal area and linearly associated is valid. As such, a higher hull painting renewal area will demand longer repainting time.

3.6.1   Repairing Time Versus Hull Coating Renewal Area (R_{TIME} vs R_{hc})

The impact of hull blasting, and painting renewal area (jointly) is also analysed. It is also assumed that the hull coating renewal area (blasting and painting area together) influences repairing time. Therefore, more hull coating renewal areas will demand more repairing time irrespective of age, deadweight and type and are expected to be linearly related.

Initial study of repairing time versus hull coating renewal area is revealed in Fig. 3.21, which shows the repairing time against the hull coating renewal area. It offers a positive linear relationship. The linear equation, $R_{\text{TIME}} = 11.272 + 0.162 * (R_{\text{hc}}/10^3)$, provides the best goodness of fit to the sample data with a correlation coefficient of 0.475.

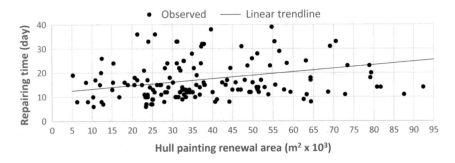

Fig. 3.20 Repairing time versus hull painting renewal area

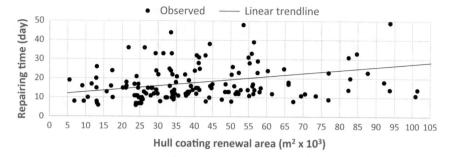

Fig. 3.21 Repairing time versus hull coating renewal area

Therefore, the assumption that the repairing time is a function of the hull coating renewal area and linearly associated is valid. Hence, a higher hull coating renewal area will demand a longer repainting time.

3.7 Repairing Time Versus Structural Steel Renewal Weight (R_{TIME} vs R_s)

It is evident that more structural steel renewal will demand more time. Therefore, the repairing time is a function of structural steel renewal weight irrespective of age, deadweight, type, and others predicted to be linearly involved.

Initial exploration of repairing time versus structural steel renewal weight is displayed in Fig. 3.22, which shows the behaviour of repairing time against structural steel renewal weight. It offers a positive linear relationship. The linear equation, $R_{TIME} = 17.221 + 0.565 * (R_s/10^3)$, yields the best goodness of fit to the sample data with a correlation coefficient of 0.289.

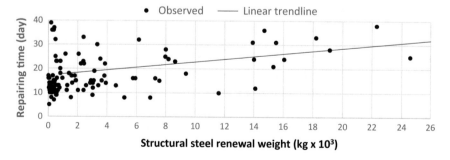

Fig. 3.22 Repairing time versus structural steel renewal weight

Therefore, the assumption that the repairing time is a function of structural steel renewal weight and linearly associated is valid. More specifically, higher structural steel renewal weight will demand longer repainting time.

3.8 Repairing Time Versus Piping Renewal Length (R_{TIME} vs R_{P})

Collected data of piping renewal covers Inert gas line on deck, tank cleaning line (COW line) on deck, fuel oil line on deck, hydraulic line on deck and in a tank, steam line (heating coil) on deck and in a tank, foam and fire line on deck, the airline on deck, sounding pipe in a tank, freshwater line, sludge & bilge line, and miscellaneous lines on deck. Other systems were not taken into consideration due to the tiny amount of repair.

The analysis considered the pipe's length only irrespective of size (diameter) and schedule (wall thickness). Pipe length is regarded as the independent variable. Therefore, the repairing time is a function of piping renewal length irrespective of age, deadweight, type, and others and is assumed to be linearly related.

An initial review of repairing time versus pipe renewal length is proven in Fig. 3.23, which shows the repairing time against piping renewal length. It offers a positive linear relationship. The linear equation, $R_{\text{TIME}} = 17.854 + 0.024 * R_{\text{p}}$, delivers the best goodness of fit to the sample data with a correlation coefficient of 0.307.

Therefore, the assumption that the repairing time is a function of piping renewal length and linearly associated is valid. More clearly, a higher pipe renewal length will demand a longer repainting time.

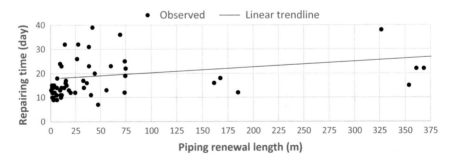

Fig. 3.23 Repairing time versus piping renewal length

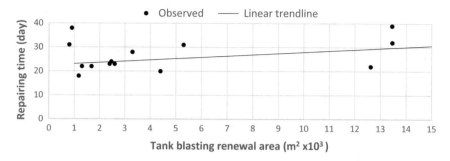

Fig. 3.24 Repairing time versus tank blasting renewal area

3.9 Repairing Time Versus Tank Blasting Renewal Area (R_{TIME} vs R_{tb})

Tank blasting works (cargo tanks and ballast tanks) are part of tank coating renewal like hull coating renewal works but with different paint quality, compatible with cargo to be carried in tanks. Blasting quality is always SA 2.5 to get a bare metal surface. After completion of blasting, the application of paint starts as per painting schemes. As such, more tank blasting renewal work will require more time. Therefore, the repairing time is a function of the tank blasting renewal area irrespective of age, deadweight, type, and others and is presumed to be linearly related.

The evaluation of repairing time versus tank blasting renewal area is exhibited in Figs. 3.24, which shows the repairing time against the tank blasting renewal area. It offers a positive linear relationship. The linear equation, $R_{\text{TIME}} = 22.566 + 0.538 * (R_{\text{tb}}/10^3)$, forecasts the best goodness of fit to the sample data with a correlation coefficient of 0.185.

Therefore, the assumption that the repairing time is a function of the tank blasting renewal area and linearly associated is valid. More particularly, a higher tank blasting renewal area will demand longer repainting time.

3.10 Repairing Time Versus Tank Painting Renewal Area (R_{TIME} vs R_{tp})

Like tank blasting, tank painting renewal work is also a part of tank coating renewal work, and more tank painting renewal works will require more time. Therefore, the repairing time is a function of tank painting renewal, irrespective of age, deadweight, type, etc., expected to have a linear relationship.

Initial investigation of repairing time versus tank painting renewal area is proved in Fig. 3.25, which shows the repairing time against tank painting renewal area. It offers a positive linear relationship. The linear equation, $R_{\text{TIME}} = 18.04 + 0.344$

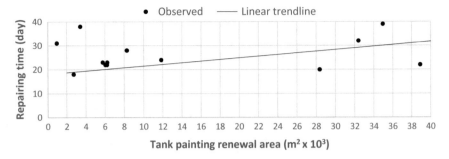

Fig. 3.25 Repairing time versus tank painting renewal area

* ($R_{tp}/10^3$), predicts the best goodness of fit to the sample data with a correlation coefficient of 0.236.

Therefore, the assumption that the repairing time is a function of the tank painting renewal area and linearly associated is valid. It means a higher tank painting renewal area will demand longer repainting time.

3.10.1 Repairing Time Versus Tank Coating Renewal Area (R_{TIME} vs R_{tc})

The impact of tank blasting and painting joint renewal area is also investigated. It is also assumed that the tank coating renewal area (blasting and painting area together) has a similar influence on repairing time. Therefore, more tank coating renewal areas will demand more repairing time irrespective of age, deadweight, and type, assuming a linear relationship.

The inspection of repairing time versus tank coating renewal area is exhibited in Fig. 3.26, which shows the repairing time against the tank coating renewal area. It offers a positive linear relationship. The linear equation, $R_{TIME} = 18.475 + 0.243$

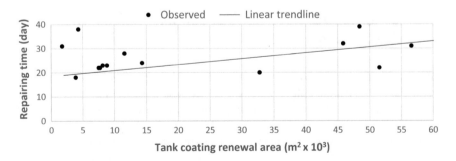

Fig. 3.26 Repairing time versus tank coating renewal area

Table 3.5 Summary of relationships and correlation coefficients under a linear form

Figure No	Variables	r^2
3.1	R_{TIME} versus S_A	0.428
3.2	R_{TIME} versus S_D	0.215
3.19	R_{TIME} versus R_{hb}	0.412
3.20	R_{TIME} versus R_{hp}	0.576
3.21	R_{TIME} versus R_{hc}	0.475
3.22	R_{TIME} versus R_S	0.289
3.23	R_{TIME} versus R_P	0.307
3.24	R_{TIME} versus R_{tb}	0.185
3.25	R_{TIME} versus R_{tp}	0.236
3.26	R_{TIME} versus R_{tc}	0.502

* $(R_{tc}/10^3)$, provides the best goodness of fit to the sample data with a correlation coefficient of 0.502.

Therefore, the assumption that the repairing time is a function of the tank coating renewal area and linearly associated is valid. So, a higher hull coating renewal area will demand a longer repainting time.

Table 3.5 summarises correlation coefficients of different relationships under linear equation forms. Based on r^2 values, it is clear from the table that the dependency of repairing time on the mentioned independent variables are at different degrees, which is anticipated. It is important to remember that the degree of dependence of the dependent variable on individual independent variables indicates its contribution to the dependent variables. Moreover, not all mentioned independent variables are involved in all routine repairing schedules. The table explains how an independent variable may influence the repairing time if that independent variable is involved.

3.11 Regression

In the previous sections, it has been highlighted that theoretically, age, deadweight, type, and renewal work like hull blasting and painting, structural steel, piping, and tank blasting and painting, are directly associated with the corresponding repairing time of a ship. In other words, repairing time (dependent variable) is a function of age, deadweight, type, renewal works (independent variables). Mathematically, the earlier relationships can be expressed in the equation form (Eqs. (3.1)–(3.11)).

$$R_{TIME} = a + b * S_A \tag{3.1}$$

$$R_{TIME} = a + b * S_D \tag{3.2}$$

$$R_{\text{TIME}} = a + b * S_{\text{T}} \tag{3.3}$$

$$R_{\text{TIME}} = a + b * R_{\text{hb}} \tag{3.4}$$

$$R_{\text{TIME}} = a + b * R_{\text{hp}} \tag{3.5}$$

$$R_{\text{TIME}} = a + b * R_{\text{hc}} \tag{3.6}$$

$$R_{\text{TIME}} = a + b * R_{\text{s}} \tag{3.7}$$

$$R_{\text{TIME}} = a + b * R_{\text{p}} \tag{3.8}$$

$$R_{\text{TIME}} = a + b * R_{\text{tb}} \tag{3.9}$$

$$R_{\text{TIME}} = a + b * R_{\text{tp}} \tag{3.10}$$

$$R_{\text{TIME}} = a + b * R_{\text{tc}} \tag{3.11}$$

Since all the independent variables are linearly associated with the dependent variable, so it is very much expected a multiple linear regression model will be an excellent fit for the system. Accordingly, a multiple linear regression model is considered to establish the relationship between repairing time, age, deadweight, renewal quantity of hull coating (blasting and painting), structural steel, piping, and tank coating (blasting and painting).

The multiple linear regression analysis is a mathematical procedure to determine the relationship involving more than one independent variable, unlike a single independent variable in simple linear regression analysis. This procedure uses the past data of both dependent and independent variables to establish the relationship and predict the dependent variable against a set of independent variables.

A general idea of the multiple linear regression analysis can be explained as follows. Let us consider an equation of the form: $y_i = b_0 + b_1 * x_{1i} + b_2 * x_{2i} + b_3 * x_{3i} \ldots b_k * x_{ki}$, where b_0, b_1, b_2, b_3, $\ldots b_k$ are the regression coefficients, k is the size of the independent variables and each set of data point [$(x_{1i}, x_{2i}, x_{3i} \ldots, x_k, y_i)$; $i = 1$, 2, 3, 4, $\ldots n$ and $n > 2$] satisfy the equation. Applying the least-squares method [8], the following normal equations can be obtained.

$$n * b_0 + b_1 \sum_{i=1}^{n} x_{1i} + b_2 \sum_{i=1}^{n} x_{2i} + \cdots + b_{ki} \sum_{i=1}^{n} x_{ki} = \sum_{i=1}^{n} y_i \tag{3.12}$$

$$b_0 \sum_{i=1}^{n} x_{1i} + b_1 \sum_{i=1}^{n} x_{1i}^2 + b_2 \sum_{i=1}^{n} x_{1i}.x_{2i} + \cdots + b_{ki} \sum_{i=1}^{n} x_{1i}.x_{ki} = \sum_{i=1}^{n} x_{1i}.y_i \quad (3.13)$$

$$b_0 \sum_{i=1}^{n} x_{2i} + b_1 \sum_{i=1}^{n} x_{1i}.x_{2i} + b_2 \sum_{i=1}^{n} x_{2i}^2 + \cdots + b_{ki} \sum_{i=1}^{n} x_{2i}.x_{ki} = \sum_{i=1}^{n} x_{2i}.y_i \quad (3.14)$$

$$b_0 \sum_{i=1}^{n} x_{ki} + b_1 \sum_{i=1}^{n} x_{ki}.x_{1i} + b_2 \sum_{i=1}^{n} x_{ki}.x_{2i} + \cdots + b_{ki} \sum_{i=1}^{n} x_{ki}^2 = \sum_{i=1}^{n} x_{ki}.y_i \quad (3.15)$$

Using the observed sets of data points, values of the above statistical notations can be calculated. Inserting those values, the above normal equations can be converted into simultaneous equations. By solving those simultaneous equations by any standard procedure, the values of regression coefficients can be estimated, and the required regression equation can be formed. Various statistical parameters are used as a reference to demonstrate the adequacy of the mathematical model considered. These are mainly (1) standard deviation, (2) coefficient of multiple determination and (3) F statistic (calculated value and critical value at 5% significance level).

Standard deviation,

$$s = \sqrt{\frac{SSE}{n - k - 1}}$$

Coefficient of multiple determination,

$$R^2 = \frac{SSR}{SST}$$

F statistic,

$$f = \frac{SSR/k}{SSE/(n - k - 1)} = \frac{SSR/k}{s^2}$$

$$f_{critical} = f_\alpha(k, n - k - 1)$$

$\alpha = 0.05$ (5% significance level)

$$SSE = SST - SSR$$

$$SST = \sum_{i=1}^{n} (y_i - \bar{y})^2 = \sum_{i=1}^{n} y_i^2 - \frac{\left(\sum_{i=1}^{n} y_i\right)^2}{n}$$

$$SSR = \sum_{j=0}^{k} b_j g_j - \frac{\left(\sum_{i=1}^{n} y_i\right)^2}{n}$$

where $g_k = \sum\limits_{i=1}^{n} x_{ki} * y_i$.

In developing the regression equation, the "stepwise regression" technique is applied using the "forward selection" method [8]. This method ensures the selection of the most influential variable from a set of variables by comparing the contribution of each variable $(R\,(\beta_j|\beta_1, \beta_2, \ldots \beta_{j-1})$ and F statistic, at every step.

To establish the relationship between the repairing time and its independent variables, the following functions (Eqs. (3.16)–(3.18)) are chosen because the repairing time is a function of each independent variable as per primarily mentioned assumptions. Equation (3.16) is designed when repairing time can be estimated using age, deadweight, and type only. Equation (3.17) is when repairing time can be calculated using age, deadweight, and designated repairing items. Finally, Eq. (3.18) is applicable for tankers (crude, chemical and product) only.

$$R_{TIME} = f(S_A, S_D, S_T) \tag{3.16}$$

$$R_{TIME} = f(S_A, S_D, R_{hb}, R_{hp}, R_s, R_p) \tag{3.17}$$

$$R_{TIME} = f(S_A, S_D, R_{hb}, R_{hp}, R_s, R_p, R_{tb}, R_{tp}) \tag{3.18}$$

Appropriate numerical values for S_T are calculated and assigned for types of ships for regression analysis. Table 3.6 displays the numerical values assigned to the types for the functional equations (Eqs. 3.16). It is used in regression analysis to form the regression Eqs. 3.19.

Using the observed data for R_{TIME}, S_A, S_D, S_T, R_{hb}, R_{hp}, R_s, R_p, R_{tb} and R_{tp}, the values of the statistical notations mentioned in the above normal equations, are calculated. They are inserted in the appropriate standard equations to convert them into simultaneous equations in regression coefficients. The solution of these simultaneous equations yields the estimate of regression coefficients. Subsequently, the statistical

Table 3.6 Numerical values for types of ships for regression analysis

Types of ships	Numerical values for S_T for Equation (3.16)
General cargo carrier	1.0000
Car carrier	1.0027
Container carrier	1.0159
Chemical tanker	1.1380
Bulk carrier	1.1694
Dredger	1.5520
Crude oil tanker	1.6472
LPG carrier	1.7060
LNG carrier	1.8600

Table 3.7 Summary of values of statistical parameters of final regression equations

Statistical parameters	Values of statistical parameters for		
	Equation (3.19)	Equation (3.20)	Equation (3.21)
Sample size (n)	590	46	13
No. of independent variable (k)	3	6	8
Significance level (α)	0.05	0.05	0.05
Standard deviation (s)	10.5	10.94	7.22
Coefficient of determination (R^2)	0.227	0.459	0.950
f -f_α	54.70	3.16	3.465

testing parameters are calculated to demonstrate the adequacy of the model and the regression equation is formed. The final regression equation passes the statistical quality test by F statistic (estimated value and critical value at 5% significance level) and coefficient of multiple determination (R^2). The final regression equations are as follows,

$$R_{\text{TIME}} = -2.148 + 0.660 * S_A + 0.0000287 * S_D + 7.463 * S_T \qquad (3.19)$$

$$R_{\text{TIME}} = -4.078 + 0.862 * S_A - 0.00012 * S_D - 0.002 * R_{\text{hb}}$$
$$+ 0.00092 * R_{\text{hp}} - 0.0000096 * R_s + 0.0146 * R_p \qquad (3.20)$$

$$R_{\text{TIME}} = -51.992 + 4.389 * S_A - 0.00044 * S_D - 0.0025 * R_{\text{hb}}$$
$$+ 0.0025 * R_{\text{hp}} - 0.00031 * R_s - 0.0445 * R_p$$
$$+ 0.0022 * R_{\text{tb}} - 0.0013 * R_{\text{tp}} \qquad (3.21)$$

The vital statistical parameters regarding final regression Eqs. (3.19)–(3.21) are given in Table 3.7. Significant improvement (higher coefficient of determination, R^2) is observed due to hull blasting and painting, structural steel renewal, piping renewal, and tank coating renewal work. Mathematically, R^2 value (in percentage) indicates the variation in the dependent variable is contributed for the variation in the independent variables, and the remaining $1 - R^2$ is called the error of estimation. Statistically, it is called an error of the sum of squares (SSE) or unexplained variation. This variation behaves in a random or unpredictable manner [9]. This also reflects the variation in the regression line. However, this estimation error is due to the absence of one or more influential independent variables (unquantifiable variables) responsible for the change in the dependent variable. The coefficient of multiple determination and the error of estimation are interrelated. For the higher value of the coefficient of multiple determination, the error of estimation will be low and vice versa. The lower value of the coefficient may also occur if the collected data suffers inconsistency. It means that when some of the independent variables have zero values. The collected datasheet indicates that some ships do not have any piping works and structural steel works and this situation lead to a lower coefficient value [2]. The condition $f > f_{0.05}$ (calculated F statistic and tabulated F statistic, respectively) suggested rejecting the null hypothesis. It may be concluded that there is a significant variation in their response (the dependent variable) due to the differences in independent variables in the postulated models.

3.12 Validation

Validation of a model is an alternative way to demonstrate the adequacy of the model's goodness of fit to the system in question. To authenticate postulated mathematical models' validation (Eqs. (3.19)–(3.21)), each model is applied to estimate repairing time for each ship and compare actual data regarding deviation (% error). The summary of the result is presented in Table 3.8. The table shows the outline of the variation of model values from the actual values. It is important to note that Eq. (3.19) considers only age, deadweight, and type of ship irrespective of repairing activities, Eq. 3.20 considers age, deadweight together with hull coating renewal area, structural steel renewal weight and piping renewal length and Eq. (3.21) (for crude oil tankers) considers age, deadweight together with hull coating renewal area, structural steel renewal weight, piping renewal length and tank coating renewal area. The improvement is significant in terms of the error level of all parameters.

The possible reasons for the highest positive and negative error (%) are described as follows. Regression analysis procedures, by nature, always pick up the mean values of independent and dependent variables (sample data) and accordingly yield the corresponding regression coefficients of independent variables to form the equations. So, any deviation in input variable(s) will yield an error in the dependent variable during the regression equation application. Later, it will be shown that if the deviation is positive, then the error will be positive and vice versa. But it will not be the case always. The regression equation dictates the relationship between deviations in input (independent variables) and output (dependent variable). Investigations on the validation result with the highest positive error (%) and highest negative error reveal some facts. For Eq. 3.19, the deviations of input data for S_A, S_D, and S_T are −65% to 248% and 2% to 63% for the highest positive and negative errors, respectively.

Table 3.8 Summary of validation results of final regression equations

Items	Equation (3.19)	Equation (3.20)	Equation (3.21)
Positive error (%)	645	153	37
Negative error (%)	−77	−65	−44
Range of error (%)	723	218	81
Mean error (%)	17	14	−2
Variance	3426	2268	617
Standard deviation	59	48	25

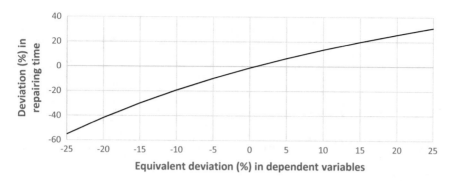

Fig. 3.27 Deviation in repairing time versus equivalent deviation in independent variables

For Eq. 3.20, the deviations of input data for S_A, S_D, R_{hb}, R_{hp}, R_s, and Rp are -82% to 87% and -93% to 31% for the highest positive and negative errors, respectively. For Eq. (3.21), the deviations of input data for S_A, S_D, R_{hb}, R_{hp}, R_s, R_p, R_{tb} and R_{tp} are -94% to 176% and -66% to 26% for the highest positive and negative errors, respectively. Investigations also reveal that observed repairing time plays a vital role in higher error (positive and negative). Their values, exceptionally high or low (for unknown reasons), also contribute to high error. For example, the mean repairing time for dredgers is 19.50 days, but a sample with 4 repairing days yields an error of 318%. The same phenomena are observed in other equations but with different magnitudes.

Typically, when the values of independent variables are closer to the sample's average value, the model will yield a reasonable estimate with negligible deviation. It can be demonstrated in Fig. 3.27. The figure shows that the equivalent mean value of the independent variables higher than the mean value of the sample data increased by the estimated repairing time, giving a higher positive deviation. Conversely, the mean value of the independent variables lesser than the mean value of the sample data resulted in a decrease in the estimated repairing time, giving a higher negative deviation. However, someone may choose to use the independent variables' relationship with the dependent variable, provided the independent variable in question significantly contributes to the total time compared to other independent variables. In other words, the variable must be the one that takes the longest time.

Furthermore, the skill of the workforce, location of repairing items, such as piping, on deck and in tanks and structural steel, in tanks and at shipside, have a strong influence on repairing time. The workforce in a shipyard, particularly the workforce allocated for ships, is of different skill levels. Hence, the time consumed will vary from ship to ship, even though the scope of repairing works is the same. For example, piping work in tanks will require more time than on deck for a similar work scope. Similarly, structural steel repair in tanks (say web frames) will consume more time than the shipside (shell plate) for an identical steel quantity. Proposed mathematical models (Eqs. (3.19)–(3.21)) are not designed to capture all these factors in estimating

Table 3.9 Summary of validation results of repairing time-age relationship

Type of ships →	All types	COT	CC	ChT	BC	LPGC	GC	CaC
Item ↓/Fig. No. →	3.1	3.5	3.6	3.7	3.8	3.9	3.10	3.11
Positive error (%)	688	230	300	449	837	121	148	139
Negative error (%)	−81	−75	−79	−76	−78	−50	−62	−65
Range of error (%)	769	306	380	525	915	170	210	204
Mean error (%)	38	38	25	16	50	22	15	7
Variance	5945	3360	3755	3810	16,665	1952	2396	2205
Standard deviation	77	58	61	62	129	44	49	47

repairing time. Therefore, for a ship with structural steel repair in tanks, the model's estimated repairing time is expected to be lower than the actual value.

Also, the high level of difference between the actual and model values may result from the weather factor, material, and spare parts availability. The weather is a natural factor that cannot predict and match the work schedule accordingly. If the weather is not favourable, the workforce will be idle, increasing the duration without output. If the spare parts are not available in due time, the idle workforce will be added to the total duration. The proposed mathematical models are not designed to capture all these factors in estimating the repairing time. Therefore, if a ship under routine repair suffers from unfavourable weather or a delay in spares supply, the model's estimated repairing time is expected to be lower than the actual value.

Validation is also applied for repairing time-age relationship under a linear form of equation (Fig. 3.1) and for different types of ships such as crude oil tankers (Fig. 3.5), container carriers (Fig. 3.6), chemical tankers (Fig. 3.7), bulk carriers (Fig. 3.8), liquified petroleum gas carriers (Fig. 3.9), general cargo carriers (Fig. 3.10) and car carriers (Fig. 3.11) using the best-fitted trend line equations. The results are presented in Table 3.9. It displays that validation result under individual type is better than under all types, in terms of the range of error, mean error and standard deviation of error except bulk carriers.

The same technique is applied for repairing the time-deadweight relationship under a linear form of equation (Fig. 3.2) and for different types of ships such as crude oil tankers (Fig. 3.12), container carriers (Fig. 3.13), chemical tankers (Fig. 3.14), bulk carriers (Fig. 3.15), liquified petroleum gas carriers (Fig. 3.16), general cargo carriers (Fig. 3.17) and car carriers (Fig. 3.18). The results are presented in Table 3.10. It displays that validation result under individual type is better than under all types, in terms of the range of error, mean error and standard deviation of error except crude oil tankers and container carriers.

Finally, the proposed mathematical models may be used to estimate repairing time for overall planning for a ship to undergo a routine maintenance program. Using the above model as a guide, shipyards may estimate the expected repairing time against an expected scope of repairing works of a ship. While using the model to estimate the repairing time, one may be aware of the error level in predicting repairing time for a set of work scope, as explained earlier. The model will estimate reasonable repair

Table 3.10 Summary of validation results of repairing time-deadweight relationship

Type of ships →	All types	COT	CC	ChT	BC	LPGC	GC	CaC
Item ↓/Fig. No. →	3.2	3.12	3.13	3.14	3.15	3.16	3.17	3.18
Positive error (%)	260	317	289	260	260	92	216	41
Negative error (%)	−86	−77	−72	−70	−89	−74	−54	−31
Range of error (%)	346	394	361	330	349	166	270	72
Mean error (%)	26	58	40	39	31	−11	33	8
Variance	3399	6928	4543	3925	5097	1748	4276	511
Standard deviation	58	83	67	63	71	42	65	23

time if the expected independent variables are close to the mean value. The estimate will be low and high for low and high value accordingly (Fig. 3.27). Also, one may consider allowing some allowance on top of the model value to accommodate the workforce's repair and skill effect. The shipyard may consider predicting the repairing time using the model for a ship, then collect actual-time data from that ship and compare with the expected value. By doing this exercise for a few ships, the shipyard may calculate the error in percentage with the actual value and use it as a guide to estimate the repairing time in the future.

3.13   General Conclusions

Through analysis, the most influential relationships of the dependent variable (repairing time) and various independent variables (age, deadweight, hull blasting and painting renewal area, structural steel renewal weight, hull piping renewal length, tank blasting and painting renewal area) are identified. Three multiple linear regression equations are developed of different combinations of independent variables. The validation technique is applied to appropriate equations to demonstrate the effectiveness. The validation results are displayed in tables. New Figures are developed for users of these findings (ship managers, shipyard personnel and anyone else) to estimate repairing time for planning and budgetary purpose. Table 3.11 displays the correlation coefficients for repairing time-age and repairing time-deadweight under linear relationship types. It provides an instant idea of how the age and deadweight may influence the expected repairing time.

Figure 3.28 is developed using the best goodness of fit relationship in Fig. 3.1. The figure shows the estimated repairing time against age irrespective of type and deadweight, and others.

Figure 3.29 is developed using the best goodness of fit relationship in Fig. 3.2. The figure shows the estimated repairing time against deadweight irrespective of type and age and others.

Figure 3.30 is developed using the best goodness of fit relationship in Figs. 3.5, 3.6, 3.7, 3.8, 3.9, 3.10 and 3.11. It shows the estimated repairing time against age

Table 3.11 Summary of correlation coefficients for repairing time-age and deadweight under a linear relationship for types

Types	Correlation coefficient (r^2)	
	$R_{TIME} = f(S_A)$	$R_{TIME} = f(S_D)$
All types	0.428	0.215
Crude oil tanker	0.332	0.134
Container carrier	0.507	0.500
Chemical tanker	0.500	0.889
Bulk carrier	0.151	0.211
LPG carrier	0.462	0.230
General cargo carrier	0.235	0.042
Car carrier	0.407	0.245

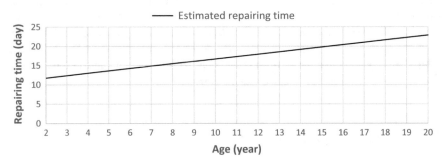

Fig. 3.28 Estimated repairing time versus age

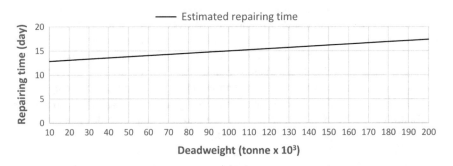

Fig. 3.29 Estimated repairing time versus deadweight

irrespective of deadweight for crude oil tankers, container carriers, chemical tankers, bulk carriers, liquified petroleum gas carriers, general cargo carriers and car carriers.

Figures 3.31 and 3.32 are developed using the best goodness of fit relationship in Figs. 3.12, 3.13, 3.14, 3.15, 3.16, 3.17 and 3.18. Figure 3.31 shows the estimated

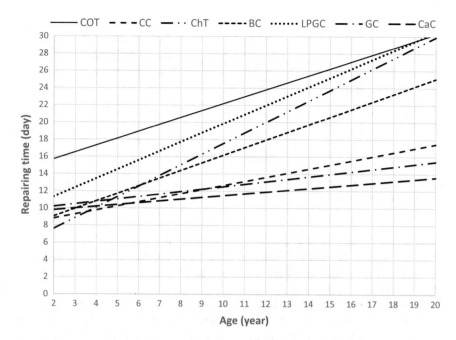

Fig. 3.30 Estimated repairing time versus age for crude oil tankers, container carriers, chemical tankers, bulk carriers, liquified petroleum gas carriers, general cargo carriers and car carriers

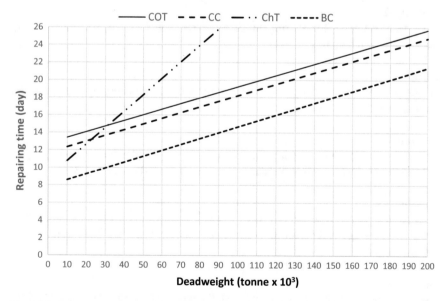

Fig. 3.31 Estimated repairing time versus deadweight for crude oil tankers, container carriers, chemical tankers, and bulk carriers

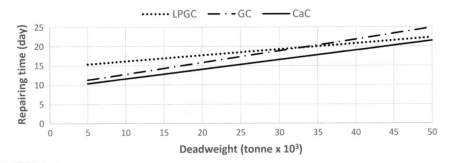

Fig. 3.32 Estimated repairing time versus deadweight for liquified petroleum gas carriers, general cargo carriers and car carriers

repairing time against deadweight irrespective of age for crude oil tankers, container carriers, chemical tankers, bulk carriers. Figure 3.32 shows the estimated repairing time against deadweight regardless of age for liquified petroleum gas carriers, general cargo carriers and car carriers.

Estimation of repairing time will vary with the available independent variables. At the preliminary stage, many variables are not available. One may like to follow the below options to estimate repairing time under the various condition of independent variables.

Option—I
Use age and estimate the repairing time irrespective of deadweight, type and repairing scopes, with the help of Fig. 3.28.

Option—II
Use deadweight and estimate the repairing time irrespective of age, type and repairing scopes, with the help of Fig. 3.29.

Option—III
Use age and type and estimate the repairing time for the corresponding type, irrespective of deadweight and repairing scopes, with the help of Fig. 3.30.

Option—IV
Use deadweight and type and estimate the repairing time for the corresponding type, irrespective of age and repairing scopes, with the help of Figs. 3.31 and 3.32 as appropriate.

Option—V
Use age, deadweight and type and estimate the repairing time for the corresponding type irrespective of repairing scopes, with regression Eq. (3.19).

Option—VI
Use age, deadweight, hull blasting and painting, structural steel renewal weight and hull piping renewal length and estimate the repairing time irrespective of type, with regression Eq. (3.20).

Option—VII
Use age, deadweight, hull blasting and painting, structural steel renewal weight, hull piping renewal length, tank blasting and painting and estimate the repairing time irrespective of type, with regression Eq. (3.21).

References

1. Apostolidis, A., Kokarakis, J., Merikas, A.: Modeling the drydocking cost—the case of tankers. J. Ship Prod. Des. **28**(3), 134–143 (2012)
2. Dev, A.K., Saha, M.: Modeling and analysis of ship repairing time. J. Ship Prod. Des. **31**(2), 129–136 (2015)
3. Dev, A.K., Saha, M.: Ship repairing time and labour. In: Proceeding, 12th Biennial International Conference, MARTECH 2017, September 20–21, 2017, Singapore (2017)
4. Surjandari, I., Novita, R.: Estimation model of dry-docking duration using data mining. World Acad. Sci. Eng. Technol. **7**(7), 1718–1721 (2013)
5. Jose, R.S.C.:. A goal programming model for vessel drydocking. J. Ship Prod. **25**(2), 95–98 (2009)
6. Emblemsvag, J.: Lean project planning in shipbuilding. J. Ship Prod. Des. **30**(2), 79–88 (2014)
7. Victoria, D., Dennis, F., Lisa, H., David, T.: Transforming the shipbuilding and ship repair project environment. J. Ship Prod. Des. **26**(4), 265–272 (2010)
8. Walpole, R.E., Myers, R.H.: Probability and Statistics for Engineers and Scientists. Macmillan Publishing Co., Inc., New York, NY (1978)
9. Murray, R.S.: Theory and Problems on Statistics. McGraw-Hill International, UK (1992)

Chapter 4
Drydocking Time

4.1 Introduction

Docking a ship refers to putting her in a graving dock (drydock), floating dock or slipway. Drydocking of a ship refers to putting her in a drydock (graving dock) only. The drydocking time may be defined as the number of days of stay in a drydock, i.e., days between dock-in and dock-out. Drydocking a ship is a routine activity that occurs regularly required by the Classification Society's rules, flag administrations', and statutory bodies' requirements. The purpose of drydocking is to carry out different surveys/inspections, particularly the underwater items. Table 4.1 demonstrates the sequence of various surveys under the timeline that a ship requires to go through regularly during her service life. Each survey has its criteria, interval, and items to be surveyed. Accordingly, ships are required to go to a drydock. Typically, the ships are needed to go to a drydock (graving dock, floating dock, or slipway) to carry out docking surveys and special surveys. It appears from Table 4.1 that for every two and a half years, a ship is required to go to a drydock for docking survey and every five years for a special survey. Other surveys, such as an annual survey and intermediate survey, are carried out at floating conditions.

It is easy to understand that the drydocking time (number of days in the drydock) is planned based on the estimated scope of drydocking works that cannot be performed afloat. Tables 3.1 and 3.2 in Chap. 3 find the docking items under routine maintenance and occasional maintenance items. This Chapter will focus on the quantifiable standard maintenance works which take place in each docking time. Table 3.1 mentioned that hull blasting, and painting works are the most common, quantifiable, and time-consuming items. Shipyards usually use the hull blasting and painting specifications provided by the shipowners to prepare the drydocking schedule and mobilise the resources. Of course, many surprises are still waiting in the drydock. After drydocking the ship and surveying, there may be some bottom damage, stern tube seal clearance beyond a limit, damage in propeller blades and so on. Widespread

Table 4.1 Surveys through the timeline

Age (year)	Event	Age (year)	Event	Age (year)	Event
1	GD & AS-1	11	AS-11	21	AS-21
2	AS-2	12	AS-12	22	AS-22
2.5	DS-1	12.5	DS-5 & IS-2	22.5	DS-9 & IS-4
3	AS-3	13	AS-13	23	AS-23
4	AS-4	14	AS-14	24	AS-24
5	AS-5, DS-2 & SS-1	15	AS-15, DS-6 & SS-3	25	AS-25, DS-10 & SS-5
6	AS-6	16	AS-16	26	AS-26
7	AS-7	17	AS-17	27	AS-27
7.5	DS-3 & IS-1	17.5	DS-7 & IS-3	27.5	DS-11, IS-5
8	AS-8	18	AS-18	28	AS-28
9	AS-9	19	AS-19	29	AS-29
10	AS-10, DS-4 & SS-2	20	AS-20, DS-8 & SS-4	30	AS-30, DS-12 & SS-6

surprise is the increase in hull blasting and painting scope of works. The most challenging part is that this type of change in no way can be known without going to the drydock.

The shipyard must adjust the plan due to the change, decrease or increase if the difference is beyond the allowable limit under the existing plan. This type of situation is seen in numerous cases, which resulted in the extension of drydocking time. This type of situation can be avoided if hull coating work is estimated more accurately. The issue will be addressed from a practical perspective by analysing the repairing time and corresponding variables.

On many occasions, the scope of repairing works was identical for two identical ships, even for the same ship in different drydocking. But the prevailing situation in the shipyard was utterly different in two different drydocking. For example, for the first drydocking, the ship was moored alongside the quay, whereas, for the second drydocking, the ship was moored alongside another ship (restricted material handling). In another scenario, the ship was moored alongside the quay before going to drydock for the first drydocking. For the second drydocking, the ship is put into drydock directly on arrival (no time for accessory works for propeller removal/shaft withdrawal) and so on.

Such a situation quickly results in a longer drydocking time. Therefore, a guideline for drydocking time for ships of various ages, deadweight, type, and hull coating renewal scope can provide a valuable reference source.

There is no documented information available about drydocking time of ships regarding their deadweight, age, and type. However, some related works were done from different viewpoints and using different variables. Dev and Saha [1] investigated ship repairing time (total days counting from the arrival at the yard to the departure from the yard). It shows that the ship repairing time (day) is linearly related to ships' age, deadweight and repairing works, namely, external hull coating, structural steel, tank coating and piping. A mathematical model was developed and proposed a multiple linear regression equation to estimate expected ship repairing time for crude oil tankers using age, deadweight, and quantity of repairing works. Dev and Saha [2] examined ship repairing labour (total man-days counting from the arrival at the yard to the departure from the yard). It shows that the ship repairing labour (man-day) is linearly related to ships' age, deadweight and repairing works, namely, external hull coating, structural steel, and piping. A mathematical model was developed and proposed a multiple linear regression equation to estimate expected ship repairing labour using age, deadweight, type, and quantity of repairing works. Dev and Saha [3] analysed drydocking time (number of days inside the drydock counting from drydock-in to drydock-out). This article attempted to demonstrate the trends of drydocking time concerning ships' deadweight, age, and type of ships. The analyses suggest that drydocking time is a function of deadweight, age, and type of a ship but at different degrees of responses. It also reveals some fundamental basis for the estimation of average drydocking time for various deadweight, age, and type. All independent variables are mostly linearly associated with the dependent variable. Jose [4] studied drydocking time and cost and used multi-criteria decision-making methods called the goal programming model to minimise drydocking time and cost. The article demonstrates the technique of the goal programming model to balance the time and cost of drydocking of a ship. Surjandari and Novita [5] explored drydocking duration using the data mining technique. It explores and identifies the relationship between drydocking time and other variables responsible for drydocking works. The authors then propose a mathematical model for the estimation of drydocking time using the CART (Classification and Regression Tree) method. This drydocking time refers to the duration that a ship stays in the dock for routine maintenance works. Naffisah et al. [6] reviewed real-life drydocking maintenance time (days) using an Artificial Neural Network technique with a backpropagation algorithm. They used 29 types of works in drydock maintenance activities as input and developed and proposed a mathematical model of drydocking maintenance time estimation.

This Chapter will discuss how the age, deadweight, type and drydocking routine maintenance works of a ship influence the drydocking time. Some assumptions are made and subsequently verified through analysing the respective variable of sample data. In this analysis, the drydocking time is considered the dependent variable, and others thought independent variables.

The general assumption is that drydocking time is a function of age, deadweight, type, (age * deadweight), hull blasting renewal area, hull painting renewal area and hull coating renewal area and they are linearly associated. In other words, age, deadweight, type, (age * deadweight), hull blasting, painting, and coating renewal area each influence drydocking time having a linear relationship. It is worth mentioning that hull blasting, and painting renewal works are the most common quantifiable activities in every routine drydocking schedule. Each of them will be discussed in the following sections.

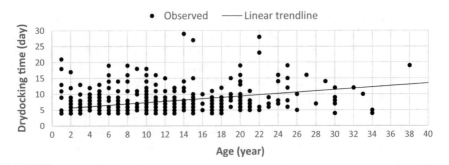

Fig. 4.1 Drydocking time versus age

4.2   Drydocking Time Versus Age (D_{TIME} vs S_A)

It is highlighted in Chap. 3 that the age of a ship has a significant impact on the repairing time. Although drydocking time is a part of the repairing time, age is also expected to impact the drydocking time positively. The relation is assumed to be linearly associated. Therefore, like repairing time, the older ship will require more time in the drydock than newer ones irrespective of deadweight, type, and others.

Initial examination of drydocking time versus age is established in Fig. 4.1, which shows the behaviour of drydocking time against age. It offers a positive linear relationship. It is likely because a ship's age dictates many drydocking items. The linear equation, $D_{TIME} = 5.328 + 0.203 * S_A$, provides the best goodness of fit to the sample data with a correlation coefficient of 0.419.

Therefore, the assumption made is valid, and older ships are expected to have a longer drydocking time than newer ones.

4.3   Drydocking Time Versus Deadweight (D_{TIME} vs S_D)

It is highlighted in Chap. 3 that the deadweight of a ship has a significant impact on the repairing time. Drydocking time being a part of the repairing time, deadweight is also likely to impact the drydocking time with a linear relationship. Therefore, like repairing time, a bigger ship will require more time in the drydock than smaller ones irrespective of age, type, and others.

Initial investigation of drydocking time versus deadweight is produced in Fig. 4.2, which shows the behaviour of drydocking time against deadweight. It offers a positive linear relationship. It is anticipated because many drydocking items are under the influence of deadweight. The linear equation, $D_{TIME} = 5.771 + 0.015 * (S_A/10^3)$, yields the best goodness of fit to the sample data with a correlation coefficient of 0.243.

Therefore, the assumption made is valid. Furthermore, it means that bigger ships are most likely to have a longer drydocking time than smaller ones.

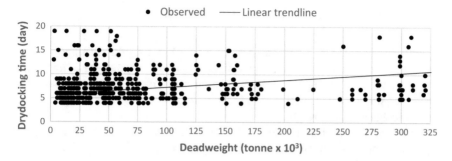

Fig. 4.2 Drydocking time versus deadweight

4.3.1 Drydocking Time Versus (age * deadweight) [D_{TIME} vs ($S_{\text{A}}*S_{\text{D}}$)]

The drydocking time is examined against the product of age and deadweight. The logic behind this is explained in Chap. 3, which is equally applicable for drydocking time. Therefore, ships with a higher (age * deadweight) value will require longer drydocking time irrespective of type.

Initial research of repairing time and (age * deadweight) is demonstrated in Fig. 4.3 showing the drydocking time against (age * deadweight) irrespective of type. The figure shows a positive relationship. Mathematically, it is pronounced because higher (age * deadweight) values require older and bigger ships, leading to a longer drydocking time. The linear equation, $D_{\text{TIME}} = 4.943 + 1.948 * [(S_{\text{A}} * S_{\text{D}})/10^6]$, delivers the best goodness of fit to sample data with a correlation coefficient of 0.784.

Therefore, the assumption made is valid. As such, bigger and older ships will require longer drydocking time than smaller and newer ships.

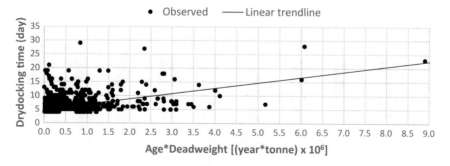

Fig. 4.3 Drydocking time versus (age * deadweight)

4.4   Drydocking Time Versus Type (D_{TIME} vs S_T)

It is discussed in Chap. 3 that the type of a ship has a significant impact on the repairing time. Drydocking time is part of the repairing time, the type of a ship will also impact the drydocking time, and the relation is assumed to be linearly associated. Therefore, different ships will require other drydocking times, even though they are similar in age and deadweight.

Initial evaluation of drydocking time versus the type of ships is introduced in Fig. 4.4. It shows a linear relationship. It is probably due to the inherent differences in the kinds of ships.

Therefore, the assumption made is valid and different ships will require other drydocking times even if they are at the same age and deadweight.

4.4.1   Drydocking Time Versus Age (D_{TIME} vs S_A) for Types

Data used in Fig. 4.4 were further analysed against age for types like crude oil tankers, container carriers, chemical tankers, bulk carriers, liquified petroleum gas carriers, general cargo carriers and car carriers. The results are presented in Figs. 4.5, 4.6, 4.7, 4.8, 4.9, 4.10 and 4.11. All these figures' essential characteristics are like that of combined types (Fig. 4.1) but with different responses.

Table 4.2 summarises the trendlines equations and correlation coefficients of the drydocking time-age relationship linearly for crude oil tankers, container carriers, chemical tankers, bulk carriers, liquified petroleum gas carriers, general cargo carriers and car carriers. It shows a lower correlation coefficient for types except for chemical tankers (Fig. 4.7) and bulk carriers (Fig. 4.8) when compared with the combined relationship of all types (Fig. 4.1). It is anticipated that the impact of age on the

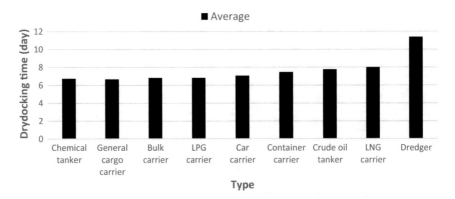

Fig. 4.4 Average drydocking time versus type

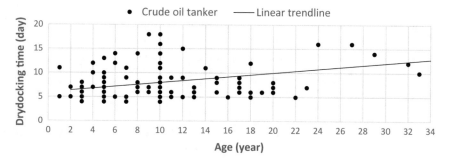

Fig. 4.5 Drydocking time versus age for crude oil tankers

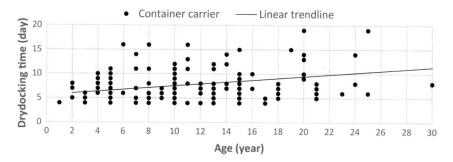

Fig. 4.6 Drydocking time versus age for container carriers

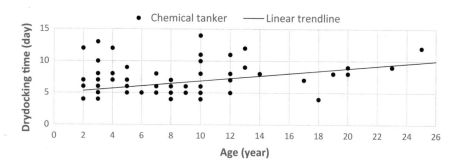

Fig. 4.7 Drydocking time versus age for chemical tankers

drydocking time widely varies for types (Table 4.2). Moreover, drydocking time varies widely for types irrespective of age and deadweight (Fig. 4.4).

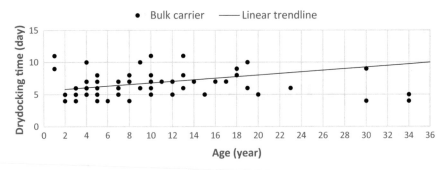

Fig. 4.8 Drydocking time versus age for bulk carriers

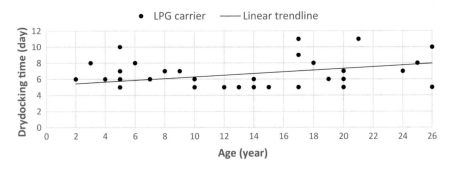

Fig. 4.9 Drydocking time versus age for liquified petroleum gas carriers

Fig. 4.10 Drydocking time versus age for general cargo carriers

4.4.2 Drydocking Time Versus Deadweight (D_{TIME} vs S_D) for Types

Data used in Fig. 4.4 were further examined against deadweight for crude oil tankers, container carriers, chemical tankers, bulk carriers, liquified petroleum gas carriers, general cargo carriers and car carriers. Results are presented in Figs. 4.12, 4.13, 4.14,

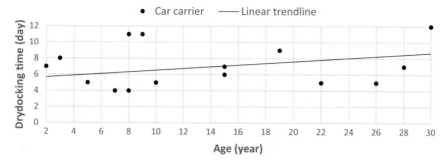

Fig. 4.11 Drydocking time versus age for car carriers

4.15, 4.16, 4.17 and 4.18. All these figures' essential characteristics are like that of combined types (Fig. 4.2) but with different responses.

Table 4.3 summarises the trendline equations and correlation coefficients of drydocking time-age relationship under a linear form for types of ships like crude oil tankers, container carriers, chemical tankers, bulk carriers, liquified petroleum gas

Table 4.2 Summary of trendline equations and correlation coefficients of drydocking time-age relationship in a linear form for types

Figure No	Trendline equations	r^2	Types
Figure 4.5	$Y = 5.984 + 0.200 * X$	0.314	Crude oil tanker
Figure 4.6	$Y = 5.622 + 0.191 * X$	0.275	Container carrier
Figure 4.7	$Y = 4.906 + 0.195 * X$	0.619	Chemical tanker
Figure 4.8	$Y = 5.568 + 0.122 * X$	0.555	Bulk carrier
Figure 4.9	$Y = 5.223 + 0.105 * X$	0.270	LPG carrier
Figure 4.10	$Y = 5.140 + 0.070 * X$	0.245	General cargo carrier
Figure 4.11	$Y = 5.453 + 0.108 * X$	0.222	Car carrier

Note X = Age (year); Y = Drydocking time (day)

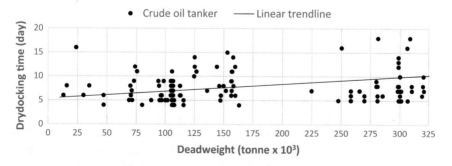

Fig. 4.12 Drydocking time versus deadweight for crude oil tankers

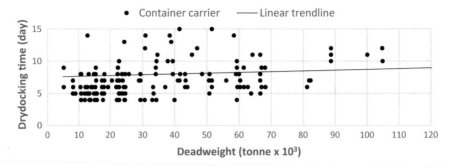

Fig. 4.13 Drydocking time versus deadweight for container carriers

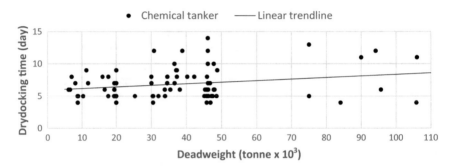

Fig. 4.14 Drydocking time versus deadweight for chemical tankers

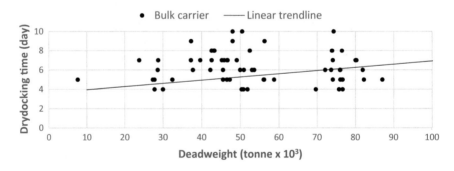

Fig. 4.15 Drydocking time versus deadweight for bulk carriers

carriers, general cargo carriers and car carriers. It shows a higher correlation coefficient except for crude oil tanker, container carrier and bulk carrier compared with combined of all types (Fig. 4.2). It is predictable because the impact of deadweight on the drydocking time widely varies for types (Table 4.3). Moreover, drydocking time varies widely for types irrespective of age and deadweight (Fig. 4.4).

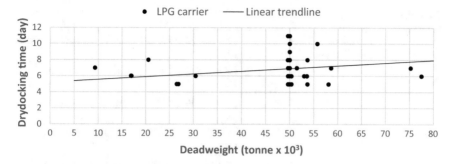

Fig. 4.16 Drydocking time versus deadweight for liquified petroleum gas carriers

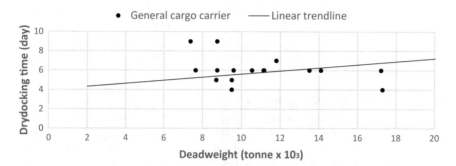

Fig. 4.17 Drydocking time versus deadweight for general cargo carriers

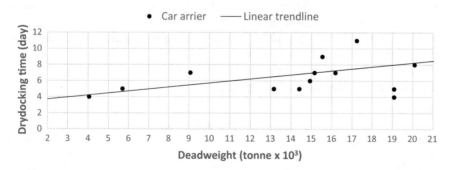

Fig. 4.18 Drydocking time versus deadweight for car carriers

4.5 Drydocking Time Versus Hull Blasting Renewal Area (D_{TIME} vs R_{hb})

It is well understood that the drydocking time is planned based on the drydocking activities, which must be completed during the drydocking time. Among the routine

Table 4.3 Summary of trendline equations and correlation coefficients of drydocking time-deadweight relationship in a linear form for types

Figure No	Trendline equations	r^2	Types
4.12	$Y = 5.372 + 0.015 * X$	0.217	Crude oil tanker
4.13	$Y = 7.547 + 0.012 * X$	0.150	Container carrier
4.14	$Y = 5.936 + 0.024 * X$	0.609	Chemical tanker
4.15	$Y = 3.623 + 0.033 * X$	0.220	Bulk carrier
4.16	$Y = 5.213 + 0.034 * X$	0.902	LPG carrier
4.17	$Y = 4.004 + 0.161 * X$	0.297	General cargo carrier
4.18	$Y = 3.211 + 0.251 * X$	0.523	Car carrier

Note $X =$ Deadweight (tonne/10^3); $Y =$ Drydocking time (day)

drydocking activities, hull blasting, and painting renewal works are every drydocking standards and significant activities. All ships go through this activity. Details of the hull coating process are elaborated in Chap. 3. Hull blasting renewal works, being a drydocking item, directly impact the drydocking time, and the relation is assumed to be linearly associated irrespective of age, deadweight, and type.

Initial examination of drydocking time versus hull blasting renewal area is demonstrated in Fig. 4.19, showing the behaviour of drydocking time against the hull blasting renewal area. It offers a positive linear relationship. It is very accurate without any doubt because more works require more time. The linear equation, $D_{TIME} = 5.851 + 0.573 * (R_{hb}/10^3)$, gives the best goodness of fit to the sample data with a correlation coefficient of 0.381.

Therefore, the assumption made is valid. As such, a higher hull blasting renewal area will demand a longer drydocking time.

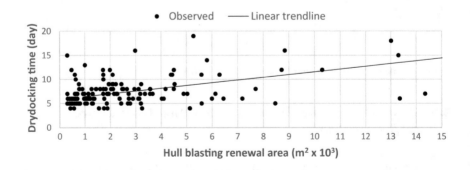

Fig. 4.19 Drydocking time versus hull blasting renewal area

4.6 Drydocking Time Versus Hull Painting Renewal Area (D_{TIME} vs R_{hp})

Like hull blasting, hull painting renewal works have a direct impact on drydocking time. Therefore, drydocking time is a function of hull painting renewal works, and the relation is assumed to be linearly associated irrespective of age, deadweight, and type.

Initial investigation of drydocking time versus hull painting renewal area is verified in Fig. 4.20, showing the drydocking time against the hull painting renewal area. It offers a positive linear relationship. It was evident because higher work volume demands a longer time. The linear equation, $D_{\text{TIME}} = 4.307 + 0.103 * (R_{\text{hp}}/10^3)$, forecasts the best goodness of fit to the sample data with a correlation coefficient of 0.510.

Therefore, the assumption is valid, and a higher hull blasting renewal area will demand a longer drydocking time.

4.6.1 Drydocking Time Versus Hull Coating Renewal Area (D_{TIME} vs R_{hc})

Like hull blasting and painting renewal works, hull coating (blasting + painting) renewal works are also studied against drydocking time. Because of the relationship between coating and blasting and painting, the coating renewal works will likely follow a linear relationship with drydocking time. Therefore, drydocking time is a function of hull coating renewal works, and the relation is assumed to be linearly associated irrespective of age, deadweight, and type.

Initial investigation of drydocking time versus hull coating renewal area is confirmed in Fig. 4.21, showing the drydocking time against the hull coating renewal area. It offers a positive linear relationship. It is very natural because coating works include blasting and painting and so the longer drydocking time. The linear equation,

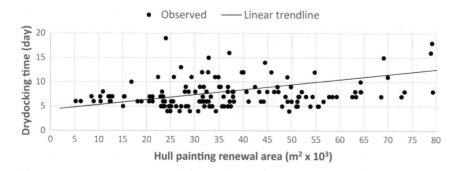

Fig. 4.20 Drydocking time versus hull painting renewal area

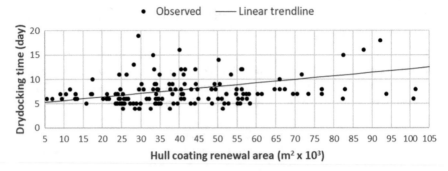

Fig. 4.21 Drydocking time versus hull coating renewal area

Table 4.4 Summary of relationships and correlation coefficients in a linear form

Figure No	Variables	r^2
4.1	D_{TIME} versus S_A	0.419
4.2	D_{TIME} versus S_D	0.243
4.19	D_{TIME} versus R_{hb}	0.381
4.20	D_{TIME} versus R_{hp}	0.510
4.21	D_{TIME} versus R_{hc}	0.367

$D_{TIME} = 4.950 + 0.072 * (R_{hc}/10^3)$, predicts the best goodness of fit to the sample data with a correlation coefficient of 0.367.

Therefore, the assumption made is valid. So, a higher hull coating renewal area will demand a longer drydocking time.

Table 4.4 summarises correlation coefficients of different relationships under linear equation forms. Based on r^2 values, it is clear from the table that the dependency of drydocking time on the mentioned independent variables are at different degrees, which is expected. It is important to remember that the degree of dependence of the dependent variable (drydocking time) on individual independent variables indicate its contribution to the dependent variables. Moreover, the combinations of independent variables are different for each routine repairing schedule. Table 4.4 gives an idea of how an independent variable may influence the drydocking time. Therefore, making a general assumption that the drydocking time is linearly associated with independent variables is not biased.

4.6.2 Relationship Between Drydocking Time and Repairing Time

In this sub-section, the focus is on how the drydocking time influences the repairing time. Drydocking time, being a part of the repairing time, undoubtedly impacts the

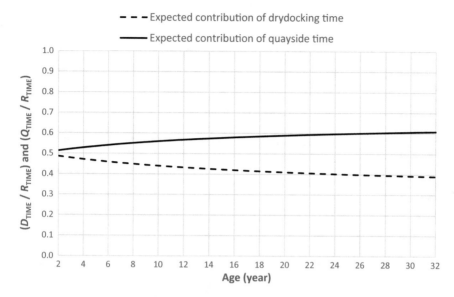

Fig. 4.22 Contribution drydocking and quayside time to repairing time versus age

repairing time. It is already established that the repairing time and drydocking time, independently, is a function of age, and those are linearly associated. Both increase with an increase of age but at a different rate. It is repairing time that increases at a higher rate than that of drydocking time. It is why it is most likely that the contribution of drydocking time to repairing time, decreases and the contribution of quayside time increases with age to satisfy the following relationship (Eq. 4.1).

$$\frac{D_{TIME}}{R_{TIME}} + \frac{Q_{TIME}}{R_{TIME}} = 1 \qquad (4.1)$$

Figures 4.22 and 4.23 theoretically demonstrate the contribution of drydocking and quayside time to repairing time versus age in fraction and percentage form, respectively. It is also observed that with the increase of age, the rate of change of contribution of drydocking time and quayside time is reduced. Mathematically, at one time, this will be so small that it may be considered negligible, and as such, the line will be almost horizontal.

Figures 4.24 and 4.25 theoretically demonstrate the contribution of drydocking time and quayside time to repairing time versus deadweight in fraction and percentage form, respectively. In this case, thing happens opposite to that of age. Here, drydocking time increases at a higher rate than that of repairing time. It is why the contribution of drydocking time to repairing time, increases and the contribution of quayside time decreases with a rise in deadweight to satisfy Eq. 4.1. Bigger ships (higher deadweight) spent very little time at the quayside than total repairing time.

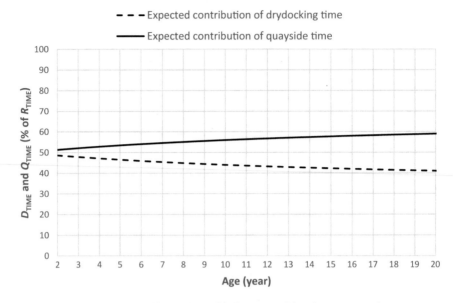

Fig. 4.23 Contribution drydocking and quayside time to repairing time versus age

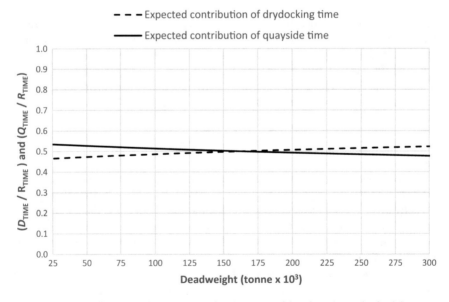

Fig. 4.24 Contribution drydocking and quayside time to repairing time versus deadweight

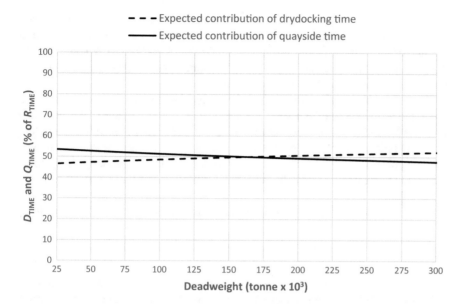

Fig. 4.25 Contribution drydocking and quayside time to repairing time versus deadweight

4.7 Regression

In the previous sections, it has been highlighted that, theoretically, age, deadweight, type and hull blasting, and painting renewal works are directly and positively associated with the corresponding drydocking time. In other words, drydocking time (dependent variable) is a function of age, deadweight, type, hull blasting renewal area and hull painting renewal area (independent variables). Mathematically, the above-mentioned relationships can be expressed in Eqs. (4.2)–(4.7).

$$D_{TIME} = a + b * S_A \tag{4.2}$$

$$D_{TIME} = a + b * S_D \tag{4.3}$$

$$D_{TIME} = a + b * S_T \tag{4.4}$$

$$D_{TIME} = a + b * R_{hb} \tag{4.5}$$

$$D_{TIME} = a + b * R_{hp} \tag{4.6}$$

$$D_{TIME} = a + b * R_{hc} \tag{4.7}$$

Since all the independent variables are linearly associated with the dependent variable, it is very likely that a multiple linear regression model will be a good fit for the system. Accordingly, a multiple linear regression model is considered to establish the relationship between drydocking time, age, deadweight, hull blasting renewal area and hull painting renewal area.

To establish the relationship between the drydocking time and its independent variables, the following functions (Eqs. 4.8 and 4.9) are chosen because the drydocking time is a function of each independent variable as per primarily mentioned assumptions. Equation (4.8) is designed when docking time can be estimated using age, deadweight, and type only. Equation (4.9) is designed when docking time can be calculated using age, deadweight, type, hull blasting and painting renewal area.

$$D_{TIME} = f(S_A, S_D, S_T) \qquad (4.8)$$

$$D_{TIME} = f(S_A, S_D, S_T, R_{hb}, R_{hp}) \qquad (4.9)$$

Appropriate numerical values for S_T are calculated and assigned for types for regression analysis. Table 4.5 displays the numerical values assigned to the types for the functional equations (Eqs. 4.8 and 4.9). It is used in regression analysis to form the regression Eqs. (4.10) and (4.11).

The final regression equations are formed following the regression analysis method mentioned in Chap. 3 and using the observed data for D_{TIME}, S_A, S_D, S_T, R_{hb}, and R_{hp}. They are as follows.

$$D_{TIME} = 0.944 + 0.117 * S_A + 6.8 \times 10^{-6} * S_D + 4.287 * S_T \qquad (4.10)$$

$$D_{TIME} = -2.84 + 0.203 * S_A + 1.49 \times 10^{-6} * S_D + 5.14 * S_T \\ + 3.75 \times 10^{-4} * R_{hb} + 2.134 \times 10^{-5} * R_{hp} \qquad (4.11)$$

Table 4.5 Numerical values for types of ships for regression analysis

Types of ships	Numerical values for S_T for	
	Equation (4.8)	Equation (4.9)
Crude oil tanker	1.6530	1.2282
Container carrier	1.1187	1.6667
Chemical tanker	1.0108	1.4167
Bulk carrier	1.0230	1.7381
LPG carrier	1.0238	1.2727
General cargo carrier	1.0000	1.0000
Car carrier	1.0600	1.1667
Dredger	1.7100	2.1667
LNG carrier	1.2000	NA

Table 4.6 Summary of values of statistical parameters of final regression equations

Statistical parameters	Values of statistical parameters for	
	Equation (4.10)	Equation (4.11)
Sample size (n)	585	151
No. of the independent variable (k)	3	5
Significance level (α)	0.05	0.05
Standard deviation (s)	3.17	3.28
Coefficient of determination (R^2)	0.098	0.348
$f - f_a$	18.31	13.21

The vital statistical parameters regarding final regression Eqs. (4.10) and (4.11) are given in Table 4.6. Significant improvement of coefficient of determination (R^2) is observed due to hull blasting, and painting renewal works. The condition, $f > f_{0.05}$ (calculated F statistic and tabulated F statistic, respectively), suggested rejecting the null hypothesis. It may be concluded that there is a significant amount of variation in their response (the dependent variable) due to the differences in independent variables in the postulated models.

4.8 Validation

The objective of validation of the proposed model is described in Chap. 3. Accordingly, the validation method is applied for Eqs. (4.10) and (4.11). The summary of the validation results is presented in Table 4.7. The table shows the outline of the variation of model values from the actual values in percentage. It is important to note that Eq. (4.10) considers only age, deadweight, and type of ship irrespective of repairing activities. Equation (4.11) assumes age, deadweight, type, hull blasting renewal area, and hull painting renewal. The improvement is significant in terms of all parameters. The possible reasons for the highest positive and negative error (%) are described as follows.

Table 4.7 Summary of validation results of final regression equations

Items	Equation (4.10)	Equation (4.11)
Positive error (%)	140	101
Negative error (%)	−73	−69
Range of error (%)	213	171
Mean error (%)	13	10
Variance	1273	1125
Standard deviation	36	34

The inherent properties of regression analysis are highlighted in Sect. 3.12 in Chap. 3. Investigations on the validation result with the highest positive error (%) and highest negative error reveal some facts. For Eq. (4.10), the deviations of input data for S_A, S_D and S_T are about $-$ 68% to 56% and $-$ 35% to 144% for the highest positive and negative errors, resulting in 140% and $-$ 73% deviation in drydocking time, respectively. For Eq. (4.11), the variations of input data for S_A, S_D, S_T, R_{hb}, and R_{hp} are about $-$ 85% to 123% and $-$ 18% to 62% for the highest positive and negative errors resulting in 101% and $-$ 69% deviation in drydocking time respectively. Other reasons for higher variations, as highlighted under repairing time analysis, are observed extraordinarily high and low values, the skill of workforces, location of works, weather, materials, and spare parts supply, are equally appropriate for drydocking time too. Figure 4.26 demonstrates the relationship of deviation (%) in docking time against equivalent variation in independent variables.

Validation is also applied for drydocking time-age relationship under a linear form of equation (Fig. 4.1) and for different types of ships such as crude oil tankers (Fig. 4.5), container carriers (Fig. 4.6), chemical tankers (Fig. 4.7), bulk carriers (Fig. 4.8), liquified petroleum gas carriers (Fig. 4.9), general cargo carriers (Fig. 4.10) and car carriers (Fig. 4.11). The results are presented in Table 4.8. It displays that the validation result under individual types is better than under all types.

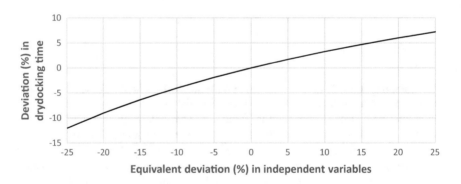

Fig. 4.26 Deviation in drydocking time versus equivalent deviation in independent variables

Table 4.8 Summary of validation results of the drydocking time-age relationship

Type of ships →	All types	COT	CC	ChT	BC	LPGC	GC	CaC
Item ↓/Fig. No. →	4.1	4.5	4.6	4.7	4.8	4.9	4.10	4.11
Positive error (%)	206	108	122	110	143	59	39	65
Negative error (%)	−74	−71	−71	−58	−70	−54	−71	−43
Range of error (%)	280	178	193	168	214	113	110	108
Mean error (%)	13	16	17	2	10	3	−3	10
Variance	1440	1393	1568	905	1254	677	729	1437
Standard deviation	38	37	40	30	35	26	27	38

Table 4.9 Summary of validation results of the drydocking time-deadweight relationship

Type of ships →	All types	COT	CC	ChT	BC	LPGC	GC	CaC
Item ↓/Fig. No. →	4.2	4.12	4.13	4.14	4.15	4.16	4.17	4.18
Positive error (%)	122	99	106	112	158	44	130	100
Negative error (%)	−77	−71	−72	−50	−70	−57	−68	−32
Range of error (%)	199	171	178	162	227	101	197	131
Mean error (%)	8	18	24	12	3	8	4	9
Variance =	1348	1667	1689	1078	1782	720	1870	1280
Standard deviation =	37	41	41	33	42	27	43	36

The same technique is applied for drydocking time-deadweight relationship under a linear form of equation (Fig. 4.2) and for different types of ships such as crude oil tankers (Fig. 4.12), container carriers (Fig. 4.13), chemical tankers (Fig. 4.14), bulk carriers (Fig. 4.15), liquified petroleum gas carriers (Fig. 4.16), general cargo carriers (Fig. 4.17) and car carriers (Fig. 4.18). The results are presented in Table 4.9. It displays that the validation result under individual type is better than under all types, except for bulk carriers and general cargo carriers.

Finally, the proposed mathematical models may be used to estimate drydocking time for a ship undergoing a routine maintenance program. Using the above model as a guide, shipyards may estimate the expected drydocking time against age, deadweight, the anticipated hull blasting and painting area scope. While using the model to estimate the scheduled docking time, one may know the error level in predicting drydocking time. The model will reasonably estimate drydocking time if the desired independent variables are close to the mean value. Accordingly, the forecast will be low and high for low and high values (Fig. 4.26). Also, one may consider allowing some allowance on top of the model value to accommodate the various variation of drydocking time explained earlier.

4.9 General Conclusions

The most influential relationships of the dependent variable (drydocking time) and various independent variables (age, deadweight, type, hull blasting and painting renewal area) are identified through analysis. Two multiple linear regression equations are developed of different combinations of independent variables. The validation technique is applied to appropriate equations to demonstrate the effectiveness. New figures are developed for users to estimate drydocking time for planning and budgetary purpose. Table 4.10 displays the correlation coefficients for drydocking time-age and drydocking time-deadweight under linear relationship types. It provides an instant idea of how the age and deadweight may influence the expected drydocking time.

Table 4.10 Summary of correlation coefficients for drydocking time-age and deadweight under a linear relationship for types

Types	Correlation coefficient (r^2)	
	$D_{TIME} = f(S_A)$	$D_{TIME} = f(S_D)$
All types	0.419	0.243
Crude oil tanker	0.314	0.217
Container carrier	0.275	0.150
Chemical tanker	0.619	0.609
Bulk carrier	0.555	0.220
LPG carrier	0.270	0.902
General cargo carrier	0.245	0.297
Car carrier	0.222	0.523

Figure 4.27 is constructed using the best goodness of fit relationship of Fig. 4.1. It demonstrates the estimated drydocking time against age irrespective of deadweight, type, and others.

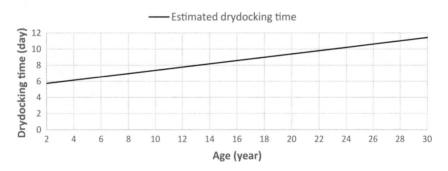

Fig. 4.27 Estimated drydocking time versus age

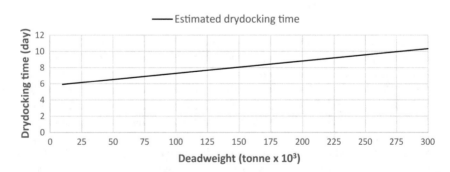

Fig. 4.28 Estimated drydocking time versus deadweight

Figure 4.28 is constructed using the best goodness of fit relationship of Fig. 4.2. It displays the estimated drydocking time against deadweight irrespective of age, type, and others.

Figures 4.29 and 4.30 are constructed using the best goodness of fit relationship in Figs. 4.5, 4.6, 4.7, 4.8, 4.9, 4.10 and 4.11. Figure 4.29 shows the estimated drydocking

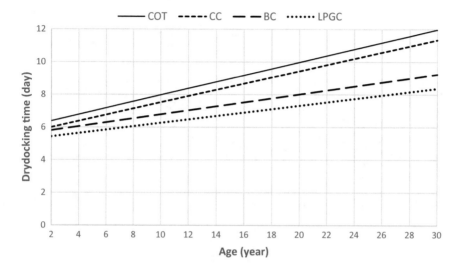

Fig. 4.29 Estimated drydocking time versus age for crude oil tankers, container carriers, bulk carriers and liquified petroleum gas carriers

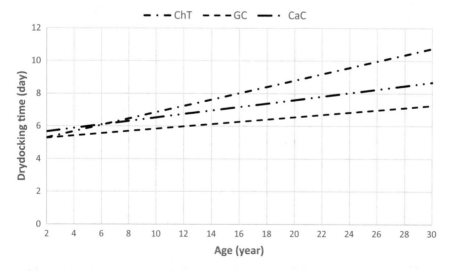

Fig. 4.30 Estimated drydocking time versus age for chemical tankers, general cargo carriers and car carriers

time against age for crude oil tankers, container carriers, bulk carriers and liquified petroleum gas carriers irrespective of deadweight.

Figure 4.30 shows the estimated drydocking time against age for chemical tankers, general cargo carriers and car carriers irrespective of deadweight.

Figures 4.31 and 4.32 are constructed using the best goodness of fit relationship in Figs. 4.12, 4.13, 4.14, 4.15, 4.16, 4.17 and 4.18. Figure 4.31 shows the estimated

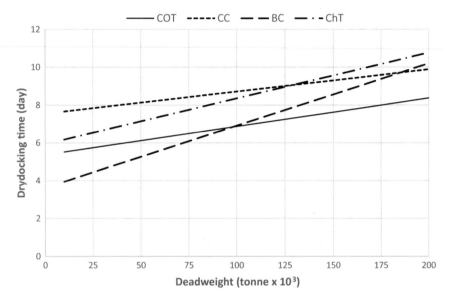

Fig. 4.31 Estimated drydocking time versus deadweight for crude oil tankers, container carriers, bulk carriers, and chemical tankers

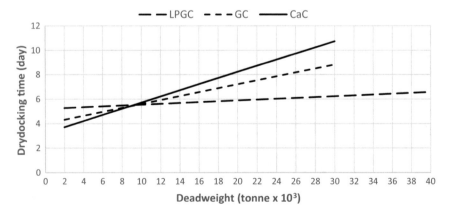

Fig. 4.32 Estimated docking time versus deadweight for liquified petroleum gas carriers, general cargo carriers and car carriers

drydocking time against deadweight for crude oil tankers, container carriers, bulk carriers, and chemical tankers irrespective of age.

Figure 4.32 shows the estimated drydocking time against deadweight for liquified petroleum gas carriers, general cargo carriers and car carriers irrespective of age.

Estimation of drydocking time will vary with the available independent variables. At the preliminary stage, many variables are not available. One may like to follow the below options to estimate drydocking time under multiple conditions of independent variables.

Option—I

Use age and estimate the drydocking time irrespective of deadweight, type, and other activities, with the help of Fig. 4.27.

Option—II

Use deadweight and estimate the drydocking time irrespective of age, type, and other activities, with the help of Fig. 4.28.

Option—III

Use age and type and estimate the drydocking time for the corresponding type, irrespective of deadweight and other activities, with the help of Figs. 4.29 and 4.30 as appropriate.

Option—IV

Use deadweight and type and estimate the drydocking time for the corresponding type, irrespective of age and other activities, with the help of Figs. 4.31 and 4.32 as appropriate.

Option—V

Use age, deadweight and type and estimate the docking time for the corresponding type irrespective of other activities, with the help of regression Eq. (4.10).

Option—VI

Use age, deadweight, type, hull blasting and painting renewal area and estimate the drydocking time with regression Eq. (4.11).

References

1. Dev, A.K., Saha, M.: Modeling and analysis of ship repairing time. J. Ship Prod. Des. **31**(2), 129–136 (2015)
2. Dev, A.K., Saha, M.: Modeling and analysis of ship repairing labor. J. Ship Prod. Des. **32**(4), 258–271 (2016)
3. Dev, A.K., Saha, M.: Dry-docking time and labour. Trans. RINA, Vol.164, Part 4. Int. J. Marit. Eng. 2018, 337–380 (2018)
4. Jose, R.S.C.: A goal programming model for vessel drydocking. J. Ship Prod. **25**(2), 95–98 (2009)

5. Surjandari, I., Novita, R.: Estimation model of dry-docking duration using data mining. World Acad. Sci. Eng. Technol. **7**(7), 1718–1721 (2013)
6. Naffisah, M.S., Surjandari, I., Rachman, A., Palupi, R.: Estimation of dry-docking maintenance duration using artificial neural network. Int. J. Comput. Commun. Instrum. Eng. (IJCCIE) **1**(1), 113–115 (2014)

Chapter 5
Ship Repairing Labour

5.1 Introduction

Repairing labour of a ship may be defined as the total workforce (man-day) utilised to complete the repairing activities during the repairing period in a shipyard, including drydocking and quayside labour. It is the summation of several men working daily onboard and in the workshop for the ship until the works are completed. The daily workforce (number of men) is planned based on work scope (quantity) and available repairing time (schedule). Logically, a shorter time needs more men per day and a longer time needs fewer men per day for the exact scope of work considering that workers are equally skilled. However, there is no clear-cut definition of shorter time and more extended time. A balance must be maintained between completing works in time, ensuring quality standards, safety, budget and optimising workforce and resources. At the end of the repair, the actual total workforce is calculated.

Chapter 3 highlighted that repairing time is a function of age, deadweight, type and repairing activities. It is known that the repairing labour is directly proportional to the repairing time. Therefore, it is evident that the repairing labour is also a function of age, deadweight, type and repairing activities.

Though sufficient literature on ship operating expenditures, ship maintenance expenditures, and related issues, very few literatures on ship repairing labour were found, the probable reason for the nonexistence of literature on ship repairing labour seems to be the non-availability of such commercial data and information strictly considered the trade secret of the concerned shipyard. However, there are some works done on ship repairing activities and related fields. Apostolidis et al. [1] evaluated and highlighted the drydocking cost for tankers. This study explores and identifies the relationship between the drydocking cost and other variables responsible for drydocking cost. The authors then propose a mathematical model for estimating the drydocking cost using the Generalised Method of Moment (GMM). This drydocking cost referred to the routine repairing cost, not the damage repairing cost. Dev and Saha [2] researched ship repairing labour (total man-days counting from arrival to

© The Author(s), under exclusive license to Springer Nature Singapore Pte Ltd. 2022
A. K. Dev et al., *Ship Repairing*, Springer Series on Naval Architecture,
Marine Engineering, Shipbuilding and Shipping 12,
https://doi.org/10.1007/978-981-16-9468-4_5

departure from the yard). It shows that the ship repairing labour (man-day) is linearly related to ships' age, deadweight and repairing works, namely, external hull coating, structural steel, and piping. A mathematical model was developed and proposed a multiple linear regression equation to estimate expected ship repairing labour using age, deadweight, type, and repairing work quantity. Dev and Saha [3] highlighted ship repairing time and labour jointly for a similar ship. This paper explores and identifies the possible independent variables responsible for ship repairing time (day) and labour (man-days). It suggested a possible relationship between various variables in a mathematical equation and verified with statistical testing parameters. A mathematical model is developed and proposed a multiple linear regression equation to estimate expected ship repairing time and labour using age, deadweight, and quantity of repairing works. Surjandari and Novita [4] demonstrated and emphasised drydocking duration using the data mining technique. This study explores and identifies the relationship between the drydocking duration and other variables responsible for drydocking works. The authors then propose a mathematical model for estimating the drydocking duration using the CART (Classification and Regression Tree) estimation method. This drydocking duration refers to the time that a ship stays in the dock for routine maintenance works. Turan et al. [5] investigated life cycle cost and earning elements and the effect of the change in structural weight due to optimisation experiments on life cycle cost and earning. They developed the life cycle cost/earning model for structural optimisation during the conceptual design stage. Meland and Spoulding [6] examined workload and labour resource planning in Northrop Grumman Newport News and provided an overview of the approach the shipyard uses to develop and manage its workload and labour resource forecasts. Naffisah et al. [7] reviewed and analysed the real-life drydocking maintenance duration (days) using the Artificial Neural Network technique with a backpropagation algorithm. They used 29 types of works in drydock maintenance activities as input and developed and proposed a mathematical model of drydocking maintenance duration estimation.

This Chapter will focus and discuss how the age, deadweight, type, and routine maintenance work scopes of a ship influence repairing labour. There will be some assumptions, and subsequently, those will be verified by analysing the repairing labour against the sample data's respective variable. In these analyses, repairing labour is considered the dependent variable, and others thought independent variables. Finally, a mathematical model is presented to calculate the repairing labour against a set of variables.

The general assumption is that repairing labour independently is a function of age, deadweight, type, (age * deadweight), hull blasting renewal area, hull painting renewal area, hull coating renewal area and structural steel renewal weight, and they are linearly associated. In other words, age, deadweight, type, (age * deadweight), hull blasting renewal area, hull painting renewal area, hull coating renewal area and structural steel renewal weight independently influence repairing labour having a linear relationship. It is essential to highlight that the above-mentioned repairing activities are the most common quantifiable events in every routine maintenance schedule. Each of them will be discussed in the following sections.

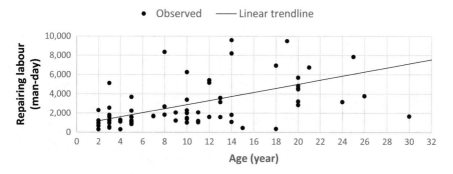

Fig. 5.1   Repairing labour versus age

5.2   Repairing Labour Versus Age (R_{LABOUR} vs S_A)

It is understood that repairing labour is directly proportional to repairing time, and logically, more repairing days means more workforce consumption. Chapter 3 shows that the repairing time is a function of age, and so, the repairing labour too is a function of age irrespective of deadweight, type, and scope of works. Therefore, the older ships are expected to consume more workforce than newer ones with a linear relationship.

The initial study of repairing labour against age is presented in Fig. 5.1, which shows repairing labour against the age irrespective of deadweight, type, and work scopes. It offers a solid and positive relationship. The linear equation, $R_{\text{LABOUR}} = 769.560 + 211.920 * S_A$, provides the best goodness of fit to the sample data with a correlation coefficient of 0.470.

This is likely because the various routine maintenance work scopes required by the classification societies rules and regulations mainly depend on the age of a ship and are reflected in the equations regarding higher labour for older ships. However, some items like hull thickness measurement during special survey and hull coating renewal work depend on the ship's physical size.

Therefore, the assumption made is valid which means, older ships will consume more workforce than newer ones.

5.3   Repairing Labour Versus Deadweight (R_{LABOUR} vs S_D)

Similarly, repairing labour is assumed to be a function of deadweight with a linear association. So, the bigger ships will consume more workforce than smaller ones.

Initial examination of repairing labour versus deadweight is depicted in Fig. 5.2, which shows the repairing labour against the deadweight irrespective of age, type, and work scopes. It offers a positive relationship. This is true because many routine

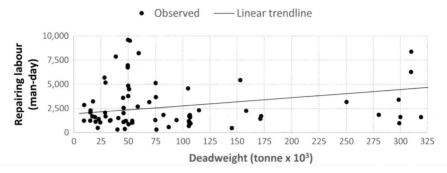

Fig. 5.2 Repairing labour versus deadweight

maintenance works depend on the ship's physical size and is reflected in the equation. The linear equation, $R_{LABOUR} = 1935.300 + 8.404 * (S_D/10^3)$, yields the best goodness of fit to the sample data with a correlation coefficient of 0.359.

Therefore, the assumption made is valid. As such, bigger ships will consume more workforce than smaller ones.

5.3.1 *Repairing Labour Versus (age * deadweight)* *[R_{LABOUR} vs (S_A * S_D)]*

Repairing labour is investigated against the product of age and deadweight (age * deadweight). The simple logic behind this is the same as explained in Chapter 3 for repairing time, which is equally applicable for repairing labour. Therefore, ships with a higher (age * deadweight) value will consume more workforce irrespective of type and are expected to be linearly associated.

Initial investigation of repairing time and (age * deadweight) is demonstrated in Fig. 5.3 showing the repairing labour against (age * deadweight) irrespective of

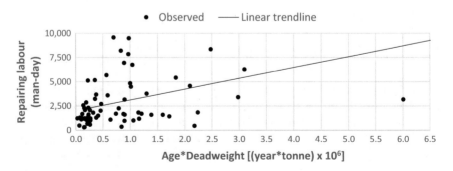

Fig. 5.3 Repairing labour versus (age * deadweight)

type. The figure shows a positive relationship. This is expected because the higher (age * deadweight) value demands bigger size and older ships, leading to more workforce consumption. The linear equations, $R_{LABOUR} = 2004.900 + 1116.300 * [(S_A * S_D)/10^6]$, delivers the best goodness of fit relationship with correlation coefficients of 0.277.

Therefore, the assumption made is valid. So, bigger and older ships will consume more labour than smaller and newer ships.

5.4 Repairing Labour Versus Type (R_{LABOUR} vs S_T)

Like repairing time, the type of a ship has a significant impact on repairing labour and is anticipated to be linearly associated. So, repairing labour is a function of type irrespective of age, deadweight, even if they are similar in size and age.

Initial examination of repairing labour versus type is confirmed in Fig. 5.4, which shows the average repairing labour against the type of ship with a strong relationship. It is possible due to the inherent differences between types of ships in terms of machinery and equipment, structural configuration, etc.

Therefore, the assumption made is valid and different ships will require different repairing labour even if they are at the same age and deadweight.

5.4.1 Repairing Labour Versus Age (R_{LABOUR} vs S_A) for Types

Data used in Fig. 5.4 were further analysed against age for types like crude oil tankers, container carriers, chemical tankers, bulk carriers and liquified petroleum gas carriers. Results are presented in Figs. 5.5, 5.6, 5.7, 5.8 and 5.9. The basic characteristics of

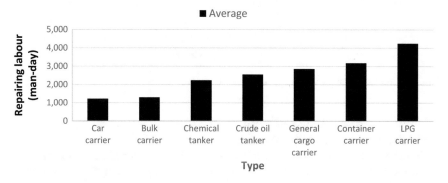

Fig. 5.4 Average repairing labour versus type

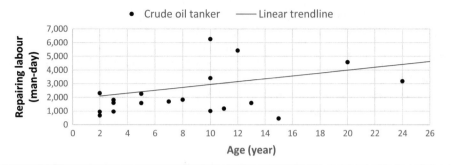

Fig. 5.5 Repairing labour versus age for crude oil tankers

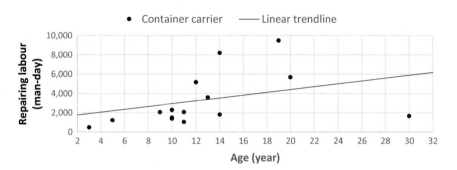

Fig. 5.6 Repairing labour versus age for container carriers

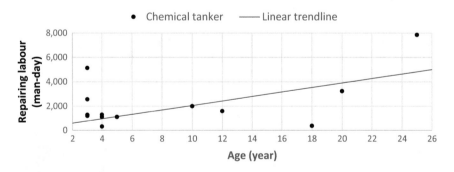

Fig. 5.7 Repairing labour versus age for chemical tankers

all these figures are like that of repairing labour against age for all types (Fig. 5.1) but with different magnitudes of response.

Table 5.1 summarises the linear trendline equations and correlation coefficients of repairing labour-age relationship for crude oil tankers, container carriers, chemical tankers, bulk carriers and liquified petroleum gas carriers. It shows lower correlation coefficients for crude oil tankers (Fig. 5.5) and container carriers (Fig. 5.6) when

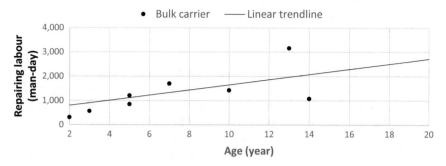

Fig. 5.8 Repairing labour versus age for bulk carriers

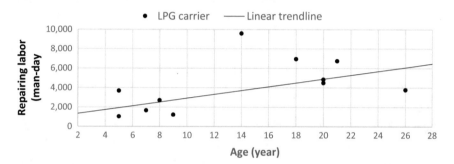

Fig. 5.9 Repairing labour versus age for liquified petroleum gas carriers

Table 5.1 Summary of linear trendline equations and correlation coefficients of repairing labour-age relationship in a linear form for types

Figure No	Trendline equations	r^2	Types
5.5	$Y = 1883.100 + 105.450 * X$	0.220	Crude oil tanker
5.6	$Y = 1470.900 + 147.180 * X$	0.170	Container carrier
5.7	$Y = 222.290 + 183.740 * X$	0.400	Chemical tanker
5.8	$Y = 588.750 + 106.250 * X$	0.375	Bulk carrier
5.9	$Y = 954.520 + 196.750 * X$	0.499	LPG carrier

Note X = Age (year), Y = Repairing labour (man-day)

compared with the combined of all types (Fig. 5.1). However, remainings are very close to that of all types.

5.4.2 Repairing Labour Versus Deadweight (R_{LABOUR} vs S_D) for Types

Data used in Fig. 5.4 were further investigated against deadweight for types like crude oil tankers, container carriers, chemical tankers, bulk carriers and liquified petroleum gas carriers. Results are presented in Figs. 5.10, 5.11, 5.12, 5.13 and 5.14. All these figures' basic characteristics are like repairing labour against deadweight for all types (Fig. 5.2) but with different magnitudes of response.

Table 5.2 summarises the linear trendline equations and correlation coefficients of repairing labour-deadweight relationship for crude oil tankers, container carriers, chemical tankers, bulk carriers and liquified petroleum gas carriers. It shows a lower correlation coefficient for container carriers (Fig. 5.11), chemical tankers (Fig. 5.12) and bulk carriers (Fig. 5.13) when compared with the combined relationship of all types (Fig. 5.2). However, remainings are very close to that of all kinds.

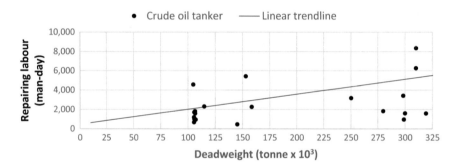

Fig. 5.10 Repairing labour versus deadweight for crude oil tankers

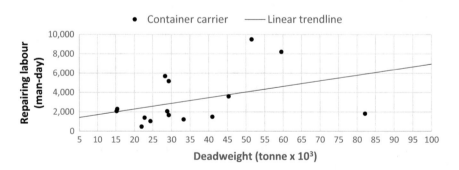

Fig. 5.11 Repairing labour versus deadweight for container carriers

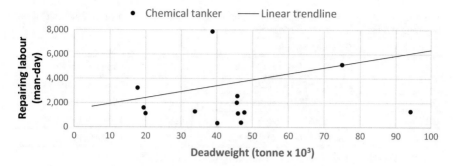

Fig. 5.12 Repairing labour versus deadweight for chemical tankers

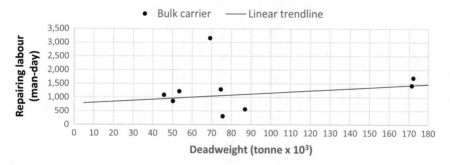

Fig. 5.13 Repairing labour versus deadweight for bulk carriers

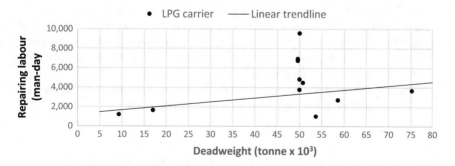

Fig. 5.14 Repairing labour versus deadweight for liquified petroleum gas carriers

5.5 Repairing Labour Versus Hull Blasting Renewal Area (R_{LABOUR} vs R_{hb})

Repairing labour is assumed to be a function of hull blasting renewal area irrespective of age, deadweight and type and linearly associated.

Table 5.2 Summary of linear trendline equations and correlation coefficients of repairing labour-deadweight relationship in a linear form for types

Figure No	Trendline equations	r^2	Types
5.10	$Y = 473.270 + 15.499 * X$	0.752	Crude oil tanker
5.11	$Y = 1128.500 + 58.204 * X$	0.149	Container carrier
5.12	$Y = 1450.000 + 48.956 * X$	0.124	Chemical tanker
5.13	$Y = 769.930 + 3.894 * X$	0.289	Bulk carrier
5.14	$Y = 1267.300 + 41.165 * X$	0.322	LPG carrier

Note X = Deadweight (tonne/10^3), Y = Repairing labour (man-day)

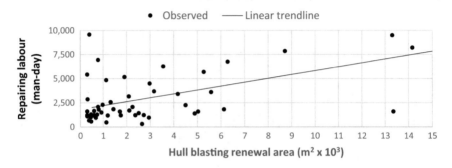

Fig. 5.15 Repairing labour versus hull blasting renewal area

Examination of repairing labour against hull blasting renewal area is demonstrated in Fig. 5.15, which shows the repairing labour against hull blasting renewal area irrespective of age, deadweight, and type. It offers a solid and positive relationship. Therefore, it is logical to expect more labour for more hull blasting renewal areas. The linear equation, $R_{LABOUR} = 1807.900 + 401.830 * (R_{hb}/10^3)$, forecasts the best goodness of fit to the sample data with a correlation coefficient of 0.593.

Therefore, the assumption made is valid. So, the hull blasting renewal area has a significant impact on repairing labour with a linear relationship.

5.6   Repairing Labour Versus Hull Painting Renewal Area (R_{LABOUR} vs R_{hp})

Repairing labour is likely to be a function of hull painting renewal area irrespective of age, deadweight and type and linearly associated.

Analysis of repairing labour against hull blasting renewal area is displayed in Fig. 5.16, which shows repairing labour against the hull painting renewal area irrespective of age, deadweight, and type. It offers a positive relationship. More hull painting renewal areas will likely demand more labour. The linear equation, R_{LABOUR}

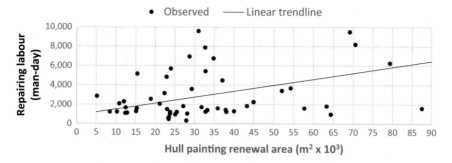

Fig. 5.16 Repairing labour versus hull painting renewal area

$= 887.590 + 61.900 * (R_{hp}/10^3)$, predicts the best goodness of fit to the sample data with a correlation coefficient of 0.460.

Therefore, the assumption made is valid. Furthermore, it means that the hull painting renewal area has a significant impact on repairing labour.

5.6.1 Repairing Labour Versus Hull Coating Renewal Area (R_{LABOUR} vs R_{hc})

The impact of the hull coating renewal area is also investigated. It is also assumed that the hull coating renewal area (blasting and painting area together) influences repairing labour with a linear relationship. Therefore, repairing labour is a function of hull coating renewal areas irrespective of age, deadweight and type and linearly associated.

Investigation of repairing labour versus hull coating renewal area is demonstrated in Fig. 5.17, which shows the pattern of repairing labour against the hull coating renewal area irrespective of age, deadweight, and type. It offers a positive linear

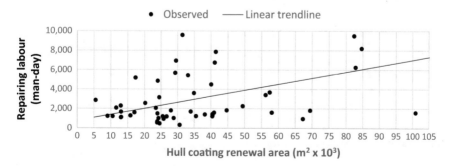

Fig. 5.17 Repairing labour versus hull coating renewal area

relationship. It is evident that more hull coating renewal areas will demand more labour. The linear equation, $R_{LABOUR} = 771.230 + 62.346 * (R_{hc}/10^3)$, provides the best goodness of fit to the sample data with a correlation coefficient of 0.556.

Therefore, the assumption made is valid. So, a higher hull coating renewal area will demand higher repainting labour.

5.7 Repairing Labour Versus Structural Steel Renewal Weight (R_{LABOUR} vs R_S)

More structural steel renewal weight will demand more workforce is very natural. Therefore, the repairing labour is a function of structural steel renewal weight irrespective of age, deadweight, type, and others and linearly involved.

Analyses of repairing labour versus structural steel renewal weight are presented in Fig. 5.18, which shows the distribution of repairing labour against structural steel renewal weight irrespective of age, deadweight, and type. It offers a solid and positive relationship. A higher structural steel renewal weight will likely demand more labour. The linear equation, $R_{LABOUR} = 2951.100 + 94.945 * (R_S/10^3)$, yields the best goodness of fit to the sample data with a correlation coefficient of 0.552.

Therefore, the assumption made is valid. As such, structural steel renewal weight has a significant impact on repairing labour.

Table 5.3 summarises correlation coefficients of various relationships under linear equation forms. It is clear from the table that the impact on repairing labour against age, deadweight, hull blasting, and painting and structural steel renewal weight are different levels of magnitudes. However, it is also important to note that the corresponding correlation coefficients seem fair based on contributors to repairing labour (independent variables). Therefore, adopting a general assumption that repairing labour is linearly associated with independent variables is not biased.

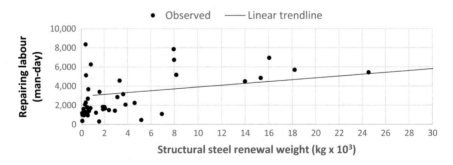

Fig. 5.18 Repairing labour versus structural steel renewal weight

Table 5.3 Summary of correlation coefficients of various relationships under a linear equation form

Figure No	Variables	r^2
5.1	R_{LABOUR} versus S_A	0.469
5.2	R_{LABOUR} versus S_D	0.359
5.15	R_{LABOUR} versus R_{hb}	0.593
5.16	R_{LABOUR} versus R_{hp}	0.460
5.17	R_{LABOUR} versus R_{hc}	0.556
5.18	R_{LABOUR} versus R_S	0.552

5.7.1   Daily Workforce and S-curve

The daily workforce is very vital for the success of a project. Too many workforces will increase the project expenditure and idle man-hours. Too little workforce will delay the completion of the project. These two extreme situations must be balanced by having proper daily workforce planning. Project duration and the estimated total workforce have a direct impact on the daily workforce mobilisation. Theoretically, a typical daily workforce curve against project time is a trapezium consisting of three sections. The first section is a straight line passing through the origin (0,0) with a positive slope indicating the gradual increase of daily workforce input. The second section is a straight line with a zero slope, indicating a steady-state condition of the daily workforce. Third, the last section is also a straight line passing through the origin (0,0) with a negative slope indicating the gradual decrease of daily workforce input. Figure 5.19 shows a typical distribution of the daily workforce against project duration. There is no hard and fast rule or formula to know or calculate the starting and end of the horizontal line. It mainly depends on the project duration. Literature does not suggest any figure for these two points, but it is understood, mathematically, the starting point will be earlier (%) for a long duration compared to a shorter-duration project.

Typically, daily workforce utilisation is monitored using the S-curve, and this S-curve is prepared using the daily workforce curve. An S-curve shows the cumulative

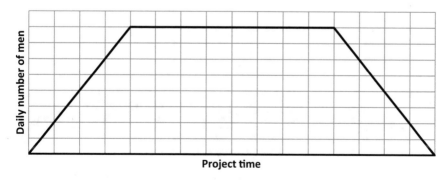

Fig. 5.19 Typical distribution of daily workforce against project time

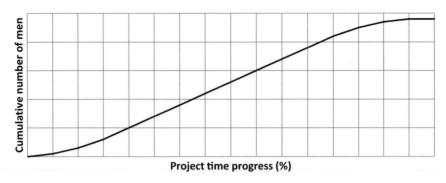

Fig. 5.20 Typical cumulative distribution of daily workforce against project time (S-curve)

workforce against the project time progress. Any point on this curve will indicate the total no. of men utilised up to that point against the progress by time, and it is compared with the actual progress of works in the field to judge if the physical progress of works is lagging or as planned or ahead of time. It is generally expected that the workforce utilisation (%) is not exceeded the job progress (%). Figure 5.20 shows a typical S-curve, the cumulative distribution of daily workforce against project time. Project time may be in days, weeks, or months, but it is shown as a percentage of project duration for comparison.

Repairing labour is studied to determine the behaviour of daily workforce (%) employed against the progressive repairing time (%), and the results is presented in Figs. 5.21 and 5.22 for observed and average daily workforce, respectively. There is a remarkable observation in Figs. 5.21 and 5.22. The final day (100%) shows a significant workforce deployment. In ship repairing it is very much obvious. This workforce is generated from the mooring gangs responsible for unberthing the ship for sailing from the shipyards and a group of technicians attending the sea trial for testing the main engine and other equipment required by the classification society's surveyor(s) and shipowner's superintendent(s). Also, in Fig. 5.22, there are no clear-cut points between which the daily workforce deployed is constant, as shown in

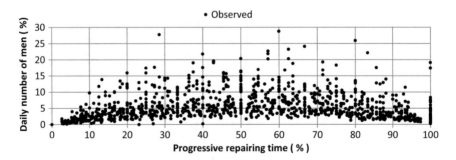

Fig. 5.21 Distribution of daily workforce against progressive repairing time

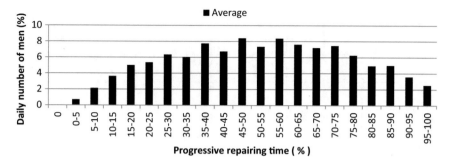

Fig. 5.22 Distribution of average daily workforce against progressive repairing time

Fig. 5.19. Figure 5.22 shows a wavy pattern. This happens due to in and out daily workforce very often. In ship repairing, many repair/maintenance items need concise time, like, anchor chain, echo sounder, sea valves, navigational equipment and are not started at the same time. Workforces responsible for these types of work usually join the workforce, finish the job, and leave the ship. An undulating nature represents this at the peak time. However, 25–80% during progressive repairing time, the daily workforce varies between 6 and 8%. This information may guide the planner for daily workforce allocation.

Same data is analysed to understand the cumulative character of the daily workforce, and results are demonstrated in Figs. 5.23 and 5.24 for the observed and average cumulative daily workforce, respectively. Essential properties are like that of Fig. 5.20 and may be used to monitor the workforce consumption and time utilised with actual work progress in the field.

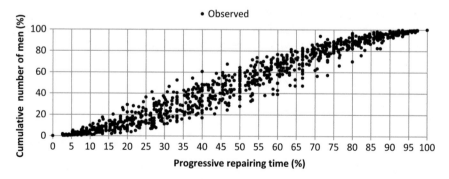

Fig. 5.23 Cumulative distribution of daily workforce against progressive repairing time (S-curve)

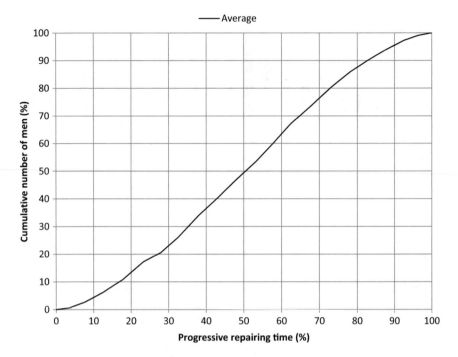

Fig. 5.24 Average cumulative distribution of daily workforce against progressive repairing time (S-curve)

5.7.2   *Daily Workforce Utilisation*

Chapter 3 discussed how the repairing time behaves against age, deadweight, type, and other independent variables. This sub-section focuses on how the daily workforce utilisation (number man per day) behaves with age, deadweight, and type.

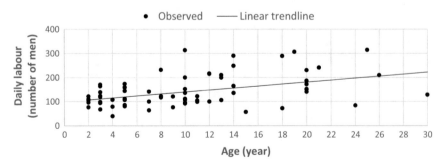

Fig. 5.25 Daily labour versus age

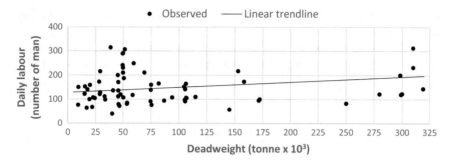

Fig. 5.26 Daily labour versus deadweight

Initial investigation of daily labour versus age is demonstrated in Fig. 5.25, which shows the behaviour of daily labour against age irrespective of deadweight and type. It offers a solid and positive relationship. The linear equation, Daily labour = 98.718 + 4.130 * S_A, delivers the best goodness of fit to the sample data with a correlation coefficient of 0.232.

Therefore, daily labour is a function of age irrespective of deadweight and type and linearly associated. So, age has a significant impact on daily labour (men per day) utilisation.

Initial investigation of daily labour versus deadweight is demonstrated in Fig. 5.26, which shows the behaviour of daily labour against deadweight irrespective of age and type. It offers a solid and positive relationship. The linear equation, Daily labour = 128.170 + 0.215 * $(S_D/10^3)$, gives the best goodness of fit to the sample data with a correlation coefficient of 0.200.

Therefore, daily labour is a function of deadweight irrespective of age and type and linearly associated. As such, deadweight has a significant impact on daily labour (men per day) utilisation.

One may wish to use the estimated repairing time and labour against age and deadweight to estimate the daily labour. In that case, Daily labour = 79.390 * $(S_A)^{0.336}$ and Daily labour = $-0.0003 * (S_D/10^3)^2 + 0.333 * (S_D/10^3) + 154.360$ with correlation coefficient of 0.998 and 0.999 respectively, will yield the best goodness of fit to the system. The above information will help to have a basic idea about the daily labour consumption. However, actual daily labour always varies with the scope of work and the time allocated for it. Usually, the daily labour utilisation pattern suggests that the daily labour during drydocking is the highest compared to other times.

5.8 Regression

So far, it has been highlighted that theoretically, age, deadweight, type, and selected renewal works like hull blasting and painting, and structural steel is directly and positively associated with the corresponding repairing labour of a ship. In other words, repairing labour (dependent variable) is a function of age, deadweight, type, renewal works (independent variables) and linearly associated. Mathematically, the above-mentioned relationships can be expressed in the equation form (Eq. 5.1–5.7).

$$R_{\text{LABOUR}} = a + b * S_A \tag{5.1}$$

$$R_{\text{LABOUR}} = a + b * S_D \tag{5.2}$$

$$R_{\text{LABOUR}} = a + b * S_T \tag{5.3}$$

$$R_{\text{LABOUR}} = a + b * R_{\text{hb}} \tag{5.4}$$

$$R_{\text{LABOUR}} = a + b * R_{\text{hp}} \tag{5.5}$$

$$R_{\text{LABOUR}} = a + b * R_{\text{hc}} \tag{5.6}$$

$$R_{\text{LABOUR}} = a + b * R_{\text{s}} \tag{5.7}$$

All the independent variables are linearly associated with the dependent variable, so a multiple linear regression model will likely be an excellent fit for the system. Accordingly, a multiple linear regression model is considered to establish the relationship between repairing labour, age, deadweight, type, renewal quantity of hull blasting, painting, and structural steel.

To establish the relationship between repairing labour and its independent variables, the following functions Eqs. (5.8) and (5.9) are chosen because repairing labour is a function of each independent variable as per primarily mentioned assumptions and analysis. Equation (5.8) is designed to estimate repairing labour using age, deadweight, and ship type only. Equation (5.9) is developed when repairing labour may be calculated using age, deadweight, and type with designated repairing items.

$$R_{\text{LABOUR}} = f(S_A, S_D, S_T) \tag{5.8}$$

$$R_{\text{LABOUR}} = f(S_A, S_D, S_T, R_{\text{hb}}, R_{\text{hp}}, R_S) \tag{5.9}$$

Appropriate numerical values for S_T are calculated and assigned for types of ships for regression analysis. Table 5.4 display the numerical values assigned to the types of ships for the functional Eqs. (5.8) and (5.9). It is used in regression analysis to form the regression Eqs. (5.10) and (5.11) respectively.

The final regression equations are formed following the regression analysis method provided in Chapter 3 and using the observed data for R_{LABOUR}, S_A, S_D,

Table 5.4 Numerical values for types of ships for regression analysis

Types of ships	Numerical values for S_T for	
	Equation (5.8)	Equation (5.9)
Bulk carrier	1.0588	1.0000
Chemical tanker	1.8365	1.3032
Crude oil tanker	2.1011	1.3410
LPG carrier	3.4887	3.1520
Container carrier	2.6147	1.9023
General cargo carrier	2.3495	1.6613
Car carrier	1.0000	NA

S_T, R_{hb}, R_{hp} and R_s. They are as follows,

$$R_{LABOUR} = -1064.700 + 142.520 * S_A + 0.005 * S_D + 837.276 * S_T \quad (5.10)$$

$$R_{LABOUR} = -1625.180 + 184.900 * S_A + 0.005 * S_D + 783.940 * S_T$$
$$- 0.020 * R_{hb} + 0.012 * R_{hp} + 0.075 * R_S \quad (5.11)$$

The vital statistical parameters regarding final regression Eqs. (5.10) and (5.11) are given in Table 5.5. Significant improvement in coefficient of multiple determination (R^2) is observed due to hull coating renewal, and structural steel renewal works. The condition $f > f_{0.05}$ (calculated F statistic and tabulated F statistic, respectively) suggested rejecting the null hypothesis. So, it may be concluded that there is a significant amount of variation in their response (the dependent variable) due to the differences in independent variables in the postulated models.

Table 5.5 Summary of values of statistical parameters of final regression equations

statistical parameters	Values of statistical parameters for	
	Equation (5.10)	Equation (5.11)
Sample size (n)	70	41
No. of the independent variable (k)	3	6
Significance level (α)	0.05	0.05
Standard deviation (s)	1958	1234
Coefficient of determination (R^2)	0.320	0.784
$f - f_\alpha$	7.58	18.16

5.9 Validation

The validation technique is applied to Eqs. (5.10) and (5.11). The summary of the validation results is presented in Table 5.6. The table shows the outline of the variation of model values from the actual values in percentage. It is important to note that Eq. (5.10) considers only age, deadweight, and type of ship irrespective of repairing activities. Equation (5.11) assumes age, deadweight, type, hull blasting and painting renewal, and structural steel renewal works. The improvement is significant in terms of all parameters. The possible reasons for the highest positive and negative error (%) are described as follows.

The inherent properties of regression analysis are highlighted in Sect. 3.12 Validation in Chapter 3. Investigations on the validation result with the highest positive error (%) and highest negative error reveal some facts. For Eq. (5.10), the deviations of input data for S_A, S_D and S_T are about -47% to 77% and -83% to -55% for the highest positive and negative errors, respectively, resulting in 89% -811% deviation in repairing labour, respectively. For Eq. (5.11), the variations of input data for S_A, S_D, S_T, R_{hb}, R_{hp} and R_S are about from -75% to 31% and -93% to -26% for the highest positive and negative errors, respectively, resulting in 140% and -76% deviation in repairing labour respectively. Other reasons for higher variations, as highlighted under repairing time analysis, like extraordinarily high and low observed values, the skill of workforces, location of works, weather, materials, and spare parts supply, are equally applicable for repairing labour too. Figure 5.27 demonstrates the relationship of deviation (%) in repairing labour against equivalent variation (%) in independent variables.

Validation is also applied for repairing labour-age relationships under the linear form of an equation for all types (Fig. 5.1) and different kinds of ships such as crude oil tankers (Fig. 5.5), container carriers (Fig. 5.6), chemical tanker (Fig. 5.7), bulk carrier (Fig. 5.8) and liquified petroleum gas carrier (Fig. 5.9). The results are presented in Table 5.7. It displays that validation result under individual type is better than under all kinds, in terms of the range of error, mean error and standard deviation of error except chemical tanker.

The same technique is applied to repairing labour-deadweight relationship under a linear form of equation (Fig. 5.2) and for different types of ships such as crude oil tankers (Fig. 5.10), container carriers (Fig. 5.11), chemical tankers (Fig. 5.12),

Table 5.6 Summary of validation results of final regression equations

Items	Equation (5.10)	Equation (5.11)
Positive error (%)	89	140
Negative error (%)	−811	−76
Range of error (%)	900	216
Mean error (%)	−56	16
Variance	21,011	2,877
Standard deviation	145	54

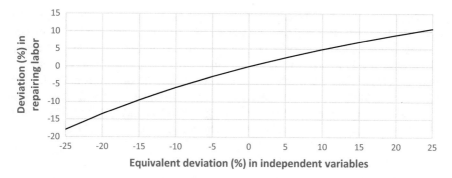

Fig. 5.27 Average deviation in repairing labour versus equivalent deviation in independent variables

Table 5.7 Summary of validation results of repairing labour-age relationship

Type of ships →	All types	COT	CC	ChT	BC	LPGC
Item ↓/Fig. No. →	5.1	5.5	5.6	5.7	5.8	5.9
Positive error (%)	1173	663	305	880	163	124
Negative error (%)	−73	−67	−57	−85	−46	−61
Range of error (%)	1246	731	362	965	209	185
Mean error (%)	76	82	71	66	28	14
Variance	33,913	24,979	12,065	65,075	4610	3455
Standard deviation	184	158	110	255	68	59

bulk carriers (Fig. 5.13) and liquified petroleum gas carriers (Fig. 5.14). The results are presented in Table 5.8. It displays that validation result under individual type is better than under all kinds, in terms of the range of error, mean error and standard deviation of error except chemical tankers.

Finally, the proposed mathematical models may estimate repairing labour for a ship undergoing a routine maintenance program. Using the above model as a guide,

Table 5.8 Summary of validation results of repairing labour-deadweight relationship

Type of ships →	All types	COT	CC	ChT	BC	LPGC
Item ↓/Fig. No. →	5.2	5.10	5.11	5.12	5.13	5.14
Positive error (%)	622	500	410	986	249	240
Negative error (%)	−75	−54	−57	−57	−67	−65
Range of error (%)	697	554	467	1043	316	306
Mean error (%)	69	106	75	239	25	10
Variance	21,052	23,061	15,771	117,002	8918	7117
Standard deviation	145	152	126	342	94	84

shipyards may calculate the expected repairing labour against age, the deadweight and anticipated scope of hull blasting area, painting area and structural steel renewal weight. While using the model to estimate the repairing labour, one may be aware of the error level in predicting repairing labour. However, if the independent variables are close to the mean value, the model reasonably estimates repairing labour. Accordingly, the estimated assessment will be low and high for low and high values (Fig. 5.27). Also, one may consider allowing some allowance on top of the model value to accommodate various reasons for repairing labour explained earlier.

5.10 General Conclusions

The most influential relationships of the dependent variable (repairing labour) and various independent variables (age, deadweight, hull blasting and painting renewal area and structural steel renewal weight) are identified. Two multiple linear regression equations are developed for different combinations of independent variables. The validation technique is applied to appropriate equations to demonstrate the effectiveness. Finally, new Figures are created for these findings to estimate repairing labour for planning and budgetary purpose.

Figure 5.28 is developed using the best goodness of fit linear equation of Fig. 5.1. The figure shows the estimated repairing labour against the age irrespective of type, deadweight, and others.

Figure 5.29 is developed using the best goodness of fit of Fig. 5.2. The figure shows the estimated repairing labour against the deadweight irrespective of type, age, and others.

Figure 5.30 is developed using the best goodness of fit of Figs. 5.5, 5.6, 5.7, 5.8 and 5.9. It shows the estimated repairing labour against the age for types like crude oil tankers, container carriers, chemical tankers, bulk carriers and liquified petroleum gas carriers.

Figure 5.31 is developed using the best goodness of fit of Figs. 5.10, 5.11, 5.12, 5.13 and 5.14. It shows the estimated repairing labour against deadweight for types

Fig. 5.28 Estimated repairing labour versus age

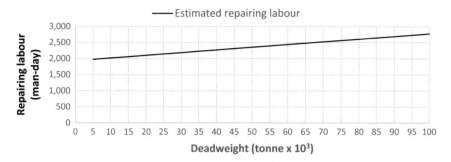

Fig. 5.29 Estimated repairing labour versus deadweight

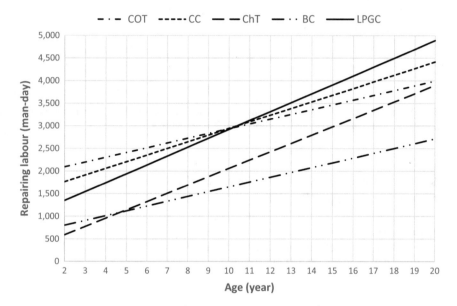

Fig. 5.30 Estimated repairing labour versus age for crude oil tankers, container carriers, chemical tankers, bulk carriers and liquified petroleum gas carriers

like crude oil tankers, container carriers, chemical tankers, bulk carriers and liquified petroleum gas carriers.

Estimation of repairing labour will vary with the available independent variables. At the preliminary stage, many variables are not available. One may like to follow the below options to estimate repairing labour under various conditions of independent variables.

Option—I

Use age and estimate the repairing labour irrespective of deadweight, type, and other activities with the help of Fig. 5.28.

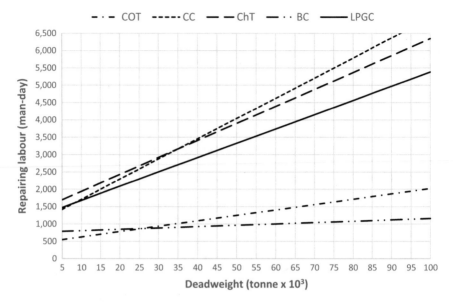

Fig. 5.31 Estimated repairing labour versus deadweight for crude oil tankers, container carriers, chemical tankers, bulk carriers and liquified petroleum gas carriers

Option—II
Use deadweight and estimate the repairing labour irrespective of age, type, and other activities with the help of Fig. 5.29.

Option—III
Use age and type and estimate the repairing labour for the corresponding type, irrespective of deadweight and other activities, with the help of Fig. 5.30.

Option—IV
Use deadweight and type and estimate the repairing labour for the corresponding type, irrespective of age and other activities, with the help of Fig. 5.31.

Option—V
Use age, deadweight and type and estimate the repairing labour for the corresponding type irrespective of other activities, with the help of regression Eq. (5.10).

Option—VI
Use age, deadweight, hull blasting and painting and structural steel renewal weight and estimate the repairing labour irrespective of other activities, with the help of regression Eq. (5.11).

References

1. Apostolidis, A., Kokarakis, J., Merikas, A.: Modeling the drydocking cost—the case of tankers. J. Ship Prod. Des. **28**(3), 134–143 (2012)
2. Dev, A.K., Saha, M.: Modeling and analysis of ship repairing labor. J. Ship Prod. Des. **32**(4), 258–271 (2016)
3. Dev, A.K., Saha, M.: Ship repairing time and labour. In: Proceeding, 12th Biennial International Conference, MARTECH 2017, 20–21 Sept. 2017, Singapore (2017)
4. Surjandari, I., Novita, R.: Estimation model of dry-docking duration using data mining. World Acad. Sci. Eng. Technol. **7** (7), 1718–1721 (2013)
5. Turan, O., Olcer, A.I., Lazakis, I., Rigo, P., Caprace, J.D.: Maintenance/repair and production-oriented life cycle cost/earning model for ship structural optimisation during the conceptual design stage. Ships Offshore Struct. **4**(2), 107–128 (2009)
6. Meland, K., Spoulding, R.: Workload and labor resource planning in a large shipyard. J. Ship Prod. **19**(1), 38–43 (2003)
7. Naffisah, M.S., Surjandari, I., Rachman, A., Palupi, R.: Estimation of dry-docking maintenance duration using artificial neural network. Int. J. Comput. Commun. Instrum. Eng. (IJCCIE) **1**(1), 113–115 (2014)

Chapter 6
Drydocking Labour

6.1 Introduction

Drydocking labour may be-defined as the workforce utilised during a drydocking period. During this time, all possible works (drydocking and non-drydocking) are carried out. Drydocking works, which cannot be done afloat, are completed during this time, and others continue until the ship departs from the yard. A detailed list of various maintenance works (drydocking and non-drydocking works) are presented in Tables 3.1 and 3.2 in Chap. 3.

Drydocking labour independently is a function of only drydocking works. Main drydocking jobs are, under routine maintenance, like hull blasting and painting, various clearance measurements like rudder pintle bushes, stern tube aft bush etc., sea valves overhauling, anchor chain calibration, chain locker cleaning and under occasional maintenance, like propeller removal, tail shaft withdrawal, tunnel thruster(s) survey and so on. In addition, rare maintenance works are carried out as per recommendations by the classification society surveyor based on survey status.

Usually, drydocking duration is planned based on drydocking items that take the longest time. For example, for routine drydocking, hull coating (blasting and painting) renewal work is the most extended time-consuming item among drydocking items. At the same time, non-drydocking works are also carried out with equal importance. Therefore, drydocking labour (labour consumed during the drydocking period) is not exclusively utilised for drydocking items only but for non-drydocking items too.

There is no documented information available about drydocking labour of ships regarding their deadweight, age, and type. However, some works, not exactly with a similar approach but close to the issue, were done from different viewpoints and using different variables. Dev and Saha [1], analysed drydocking labour (number of man-days utilised during drydocking time). This article attempts to demonstrate the trends of drydocking labour regarding ships' deadweight, age, and type of ships. The analyses suggest that drydocking labour is a function of deadweight, age, and type

A. K. Dev et al., *Ship Repairing*, Springer Series on Naval Architecture, Marine Engineering, Shipbuilding and Shipping 12, https://doi.org/10.1007/978-981-16-9468-4_6

of a ship but at different degrees of responses. It also reveals some fundamental basis for estimating average drydocking labour for various deadweight, age, and type. All independent variables are mostly linearly associated with the dependent variable. Dev and Saha [2] studied ship repairing labour (total man-days counting from the arrival at the yard to the departure from the yard). It shows that the ship repairing labour (man-day) is linearly related to ships' age, deadweight and repairing works, namely, external hull coating, structural steel, and piping. A mathematical model was developed and proposed a multiple linear regression equation to estimate expected ship repairing labour using age, deadweight, type, and quantity of repairing works.

In this Chapter, the focus is to analyse and discuss how the age, deadweight, type and drydocking routine maintenance works of a ship influence the drydocking labour. There will be some assumptions, and subsequently, it will be verified by analysing the respective sample data variable. In this analysis, drydocking labour is considered the dependent variable, and others thought independent variables.

As such, a general assumption is made that the drydocking labour is a function of age, deadweight, (age * deadweight), type, hull blasting renewal area, hull painting renewal area and hull coating renewal area independently and linearly associated. Above mentioned similar variables are considered for analysis of drydocking time (Chap. 4), too. The following sections will describe each of them.

6.2   Drydocking Labour Versus Age (D_{LABOUR} vs S_A)

As per assumption, drydocking labour is a function of age irrespective of deadweight, type, and other linearly associated activities. In other words, the older ships will consume more workforce than the newer ones in a drydock.

Investigation of drydocking labour against age is presented in Fig. 6.1, which shows drydocking labour against age irrespective of deadweight, type, and work scopes. It offers a solid and positive relationship. It is expected because the various routine maintenance work scopes required by the classification societies rules and regulations depend on the ship's age. It is reflected in the trendline in terms of higher

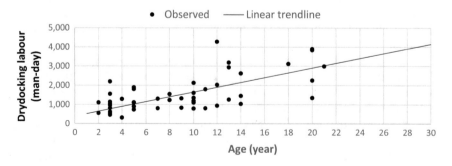

Fig. 6.1   Drydocking labour versus age

labour for a higher age. However, some items like hull thickness measurement during particular survey and hull coating renewal work depend mainly on the ship's physical size. The linear equation, $D_{LABOUR} = 398.550 + 125.150 * S_A$, provides the best goodness of fit to the sample data with a correlation coefficient of 0.830.

Therefore, the assumption made is valid. So, older ships are likely to consume more workforce in drydock than newer ones. The above findings are in line with the requirements of rules of the classification societies. Furthermore, the robustness of inspections/surveys and tests of items increases with age. As such, the age of a ship is a crucial variable for drydocking labour.

6.3 Drydocking Labour Versus Deadweight (D_{LABOUR} vs S_D)

It is assumed that the drydocking labour is a function of deadweight and linearly associated. In other words, bigger ships are expected to consume more workforce in a drydock than smaller ones irrespective of age, type, and other activities.

Initial examination of drydocking labour against deadweight is demonstrated in Fig. 6.2 showing drydocking labour against deadweight irrespective of age, type, and work scopes. It offers a positive linear relationship. It is likely because the various routine maintenance work scopes required by the CS's rules and regulations depend on the ship's physical size. It is reflected in the trendline in terms of higher labour for a higher deadweight. The linear equation, $D_{LABOUR} = 1216.600 + 1.596 * (S_D/10^3)$, provides the best goodness of fit to the sample data with a correlation coefficient of 0.084.

Therefore, the assumption made is valid but weak. As such, bigger ships will not necessarily consume more workforce in drydock than smaller ones. So, the deadweight of a ship is not an essential variable related to drydocking labour.

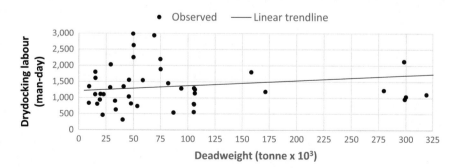

Fig. 6.2 Drydocking labour versus deadweight

6.3.1   Drydocking Labour Versus (age * deadweight), [D_{LABOUR} vs (S_A*S_D)]

The variable (age * deadweight) will likely significantly impact drydocking labour with a linear relationship. Therefore, drydocking labour is a function of (age * deadweight) and is linearly associated irrespective of type and other activities.

Initial investigation of drydocking labour versus (age * deadweight) is displayed Fig. 6.3, showing drydocking labour against (age * deadweight) irrespective of type and work scopes. It offers a strong positive relationship. It is normal because a higher (age * deadweight) value requires older and bigger ships leading to higher labour consumption. The linear equation, $D_{LABOUR} = 830.140 + 1411.800 * [(S_A*S_D)/10^6]$, provides the best goodness of fit to the sample data with a correlation coefficient of 0.852.

Therefore, the assumption made is valid. More particularly, a ship with a higher (age * deadweight) value will consume more workforce in drydock than a ship with a lower (age * deadweight) value.

6.4   Drydocking Labour Versus Type (D_{LABOUR} vs S_T)

As per initial assumptions, the type of ship has a significant impact on the drydocking labour. Therefore, the drydocking labour is a function of the type of ship irrespective of age, deadweight, even if they are similar in size and age.

Analysis of drydocking labour versus type is depicted in Fig. 6.4, which shows the behaviour of average drydocking labour against type irrespective of age, deadweight, and work scopes. It offers a strong relationship. It is very much likely because of inherent differences between types of ships. Therefore, the assumption made is valid. Due to their inherent differences, different ship types will consume an extra amount of labour, even if those are of the same age and size.

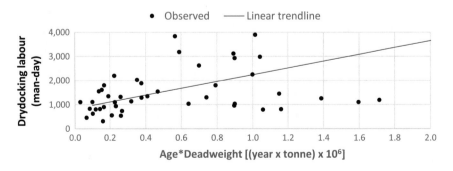

Fig. 6.3 Drydocking labour versus (age * deadweight)

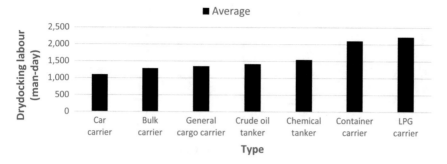

Fig. 6.4 Average drydocking labour versus type

6.4.1 Drydocking Labour Versus Age (D_{LABOUR} vs S_A) for Types

Data used in Fig. 6.4 were further investigated against age for types of ships like crude oil tankers, container carriers, chemical tankers, bulk carriers and liquified petroleum gas carriers. The results are presented in Figs. 6.5, 6.6, 6.7, 6.8 and 6.9. All these figures' basic characteristics are like drydocking labour against age for combined types (Fig. 6.1) but with different response magnitudes.

Table 6.1 summarises the linear trendline equations and corresponding correlation coefficients of drydocking labour-age relationships for crude oil tankers, container carriers, chemical tankers, bulk carriers and liquified petroleum gas carriers. It shows a lower correlation coefficient when compared with the combined of all types (Fig. 6.1).

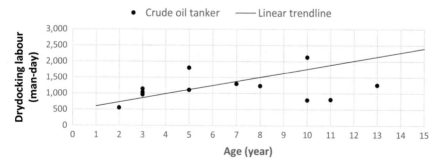

Fig. 6.5 Drydocking labour versus age for crude oil tankers

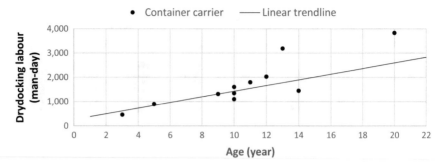

Fig. 6.6   Drydocking labour versus age for container carriers

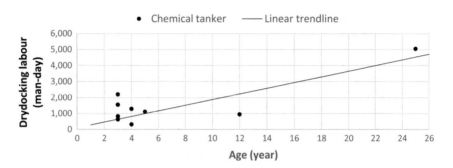

Fig. 6.7   Drydocking labour versus age for chemical tankers

Fig. 6.8   Drydocking labour versus age for bulk carriers

6.4.2   *Drydocking Labour Versus Deadweight (D_{LABOUR} vs S_D) for Types*

Data used in Fig. 6.4 were further investigated against deadweight for types of ships like crude oil tankers, container carriers, chemical tankers, bulk carriers and liquified petroleum gas carriers. Results are presented in Figs. 6.10, 6.11, 6.12, 6.13 and 6.14.

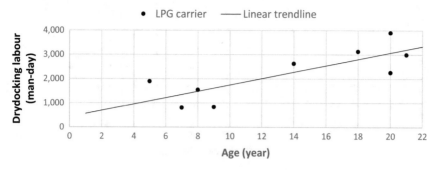

Fig. 6.9 Drydocking labour versus age for liquified petroleum gas carriers

Table 6.1 Summary of linear trendline equations and correlation coefficients of drydocking labour-age relationship in a linear form for types

Figure No	Trendline equations	r^2	Types
6.5	$Y = 475.960 + 128.190 * X$	0.217	Crude oil tanker
6.6	$Y = 265.420 + 116.460 * X$	0.743	Container carrier
6.7	$Y = 105.890 + 176.820 * X$	0.817	Chemical tanker
6.8	$Y = 145.640 + 126.860 * X$	0.415	Bulk carrier
6.9	$Y = 426.630 + 132.000 * X$	0.727	LPG carrier

Note X = Age (year); Y = Drydocking labour (man-day)

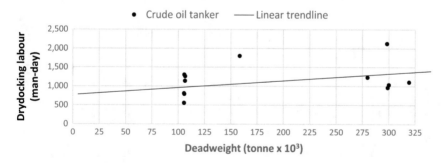

Fig. 6.10 Drydocking labour versus deadweight for crude oil tankers

All these figures' basic characteristics are like drydocking labour against deadweight for combined types (Fig. 6.2) but with different response magnitudes.

Table 6.2 summarises the linear trendline equations and corresponding correlation coefficients of drydocking labour-deadweight relationships for crude oil tankers, container carriers, chemical tankers, bulk carriers and liquified petroleum gas carriers. It shows a higher correlation coefficient for types when compared with that of combined types (Fig. 6.2).

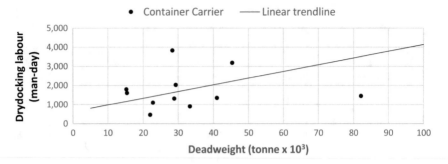

Fig. 6.11 Drydocking labour versus deadweight for container carriers

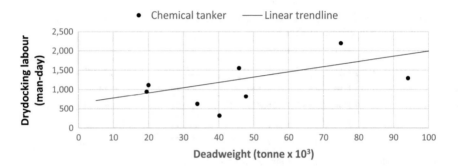

Fig. 6.12 Drydocking labour versus deadweight for chemical tankers

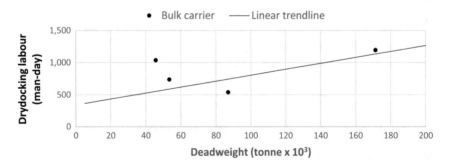

Fig. 6.13 Drydocking labour versus deadweight for bulk carriers

6.5 Drydocking Labour Versus Hull Blasting Renewal Area (D_{LABOUR} vs R_{hb})

As per the initial assumption, the drydocking labour is a function hull blasting renewal area and is linearly associated.

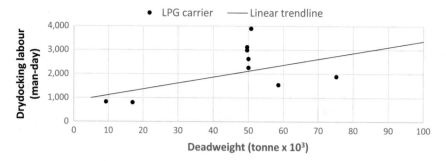

Fig. 6.14 Drydocking labour versus deadweight for LPG carriers

Initial examinations of drydocking labour against hull blasting renewal area presented in Fig. 6.15 show drydocking labour against hull blasting renewal area irrespective of age, deadweight, and type. It offers a solid and positive relationship. It is most likely, as pointed out in Sect. 6.2. The linear equation, $D_{LABOUR} = 1041.100 + 254.980 * (R_{hb}/10^3)$, provides the best goodness of fit to the sample data with a correlation coefficient of 0.608.

Table 6.2 Summary of linear trendline equations and correlation coefficients of drydocking labour-deadweight relationship in a linear form for types

Figure No	Trendline equations	r^2	Types
6.10	$Y = 773.620 + 1.840 * X$	0.929	Crude oil tanker
6.11	$Y = 628.910 + 35.126 * X$	0.183	Container carrier
6.12	$Y = 645.640 + 11.477 * X$	0.361	Chemical tanker
6.13	$Y = 343.710 + 4.600 * X$	0.700	Bulk carrier
6.14	$Y = 869.970 + 24.899 * X$	0.277	LPG carrier

Note X = Deadweight (tonne/10^3); Y = Drydocking labour (man-day)

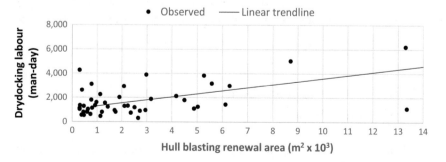

Fig. 6.15 Drydocking labour versus hull blasting renewal area

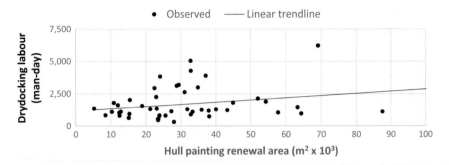

Fig. 6.16 Drydocking labour versus hull painting renewal area

Therefore, the assumption made is valid. Hence, the hull blasting renewal area significantly influences the drydocking labour, and it is likely because it is purely a major drydocking item.

6.6 Drydocking Labour Versus Hull Painting Renewal Area (D_{LABOUR} vs R_{hp})

Like the hull blasting renewal area, drydocking labour is also a function of the hull painting renewal area and is linearly associated.

Initial investigations of drydocking labour versus hull painting renewal area demonstrated in Fig. 6.16 show drydocking labour against the hull painting renewal area irrespective of age, deadweight, and type. It offers a positive relationship. It is very much likely, as pointed out in Sect. 6.3. The linear equation, $D_{\text{LABOUR}} = 1169.400 + 16.970 * (R_{\text{hp}}/10^3)$, provides the best goodness of fit to the sample data with a correlation coefficient of 0.090.

Therefore, the assumption made is valid but weak. As such, the hull painting renewal area has a weak influence on drydocking labour. The real-life scenario also suggests similar phenomena because blasting processes consume more labour than the painting process.

6.7 Drydocking Labour Versus Hull Coating Renewal Area (D_{LABOUR} vs R_{hc})

Like hull blasting and painting renewal work, hull coating (blasting and painting area together) renewal work also influences drydocking labour. Therefore, drydocking labour is a function of hull coating renewal area irrespective of age, deadweight, and type.

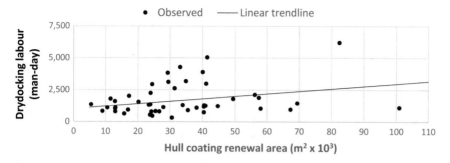

Fig. 6.17 Drydocking labour versus hull coating renewal area

Analyses of drydocking labour versus hull coating renewal area displayed in Fig. 6.17 show the behaviour of drydocking labour against hull coating renewal area irrespective of age, deadweight, and type. It offers a positive relationship. It is expected because of the combined effect of blasting and painting. The linear equation, $D_{LABOUR} = 105.420 + 53.310 * (R_{hc}/10^3)$, provides the best goodness of fit to the sample data with a correlation coefficient of 0.625.

Therefore, the assumption is valid. Hence, the hull coating renewal area influences the drydocking labour. It is very likely because it is the longest time consuming and highest labour-consuming among the activities during routine drydocking.

6.7.1 Relationship Between Drydocking Labour and Repairing Labour

In this sub-section, the focus is directed to how drydocking labour influences the repairing labour, in the same way drydocking time influences repairing time, in Chap. 4. Drydocking labour, being a part of the repairing labour, undoubtedly impacts the repairing labour up to some extent. It is already established that the repairing and drydocking labour, independently, is a function of age, deadweight, and linearly associated. Both increase with an increase of age but at a different rate. Repairing labour increases at a higher rate than that drydocking labour. It is expected that the contribution of drydocking labour to repairing labour, increases and the contribution of quayside labour decreases with an increase of age to satisfy the below relationship (Eq. 6.1).

$$\frac{D_{LABOUR}}{R_{LABOUR}} + \frac{Q_{LABOUR}}{R_{LABOUR}} = 1 \tag{6.1}$$

Theoretically, Figs. 6.18 and 6.19 demonstrate the contribution of drydocking labour and quayside labour to repairing labour versus age, in fraction and percentage form, respectively. It is also observed that the rate of change of contribution of

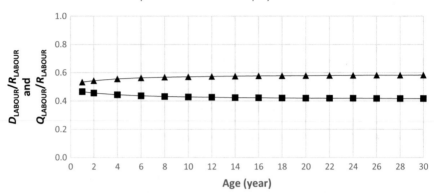

Fig. 6.18 Contribution of drydocking and quayside labour to repairing labour versus age

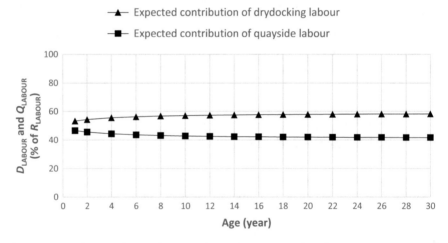

Fig. 6.19 Contribution of drydocking and quayside labour to repairing labour versus age

drydocking labour and quayside labour is reduced with the increase of age. Mathematically, this change after a certain age is so small that it may be considered negligible (if not zero), and as such, the line will be almost horizontal.

Theoretically, Figs. 6.20 and 6.21 demonstrate the contribution of drydocking labour and quayside labour to repairing labour versus deadweight in fraction and percentage form, respectively. The contribution of drydocking labour decreases with the increased deadweight, and obviously, the contribution of quayside labour increases to satisfy Eq. (6.1).

Above Figs. 6.18, 6.19, 6.20 and 6.21 display the behaviour of contribution of drydocking and quayside labour to repairing labour (as fraction and percentage of

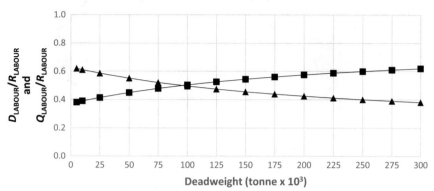

Fig. 6.20 Contribution of drydocking and quayside labour to repairing labour versus deadweight

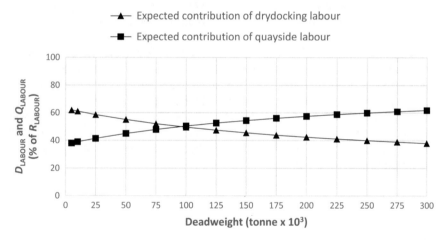

Fig. 6.21 Contribution of drydocking and quayside labour to repairing labour versus deadweight

repairing labour) against age and deadweight. Now the focus will be on the daily workforce (number of men) deployed during drydocking and along the quayside. This simple investigation aims to find the breakdown of the total repairing labour and daily workforce used during drydocking time and during alongside irrespective of age, deadweight, type, and work scopes. There is no hard and fast rule to split the repairing labour into drydocking and quayside labour. It depends on the type of works (drydocking items like a significant amount of bottom plate renewal and non-drydocking like tank coating etc.). Still, there is always an expectation and sincere attempt to reduce the drydocking time to save cost (for owners) and increase revenue (for shipyards). First, however, collected data of repairing labour are examined, and

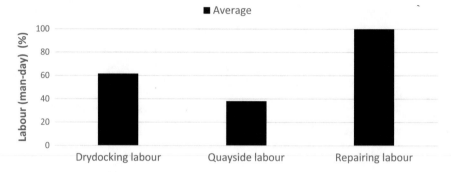

Fig. 6.22 Average breakdown of repairing labour

the result is presented in Fig. 6.22. The figure shows that 60% of total labour was utilised during drydocking for sample ships on average. This information may serve the planner as a guideline to plan.

Daily workforce allocation in drydock and quayside also totally depends on the type of works. But there is always an urge to engage more workforce in drydock to finish related works with a noble objective to reduce drydocking time. However, collected data of repairing labour are explored, and the result is exhibited in Fig. 6.23. The figure shows that on average daily, 223 and 112 men were utilised in drydock and quayside respectively for sample ships. Whereas overall average, every day 169 men were employed. This finding is in line with the finding in Fig. 6.22 that the labour consumption in drydock is higher than at quayside. Thus, once again, the total and daily workforce during drydocking is higher than that of quayside. This information also surfaces a challenge for planners to think and find ways to reduce workforce consumption and repair costs.

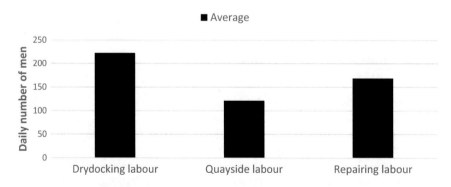

Fig. 6.23 Average daily number of men in a drydock and a quayside

6.8 Regression

In the previous sections, it has been highlighted that theoretically, age, deadweight, type, and selected renewal works like hull blasting and painting are directly and positively associated with the corresponding drydocking labour. In other words, drydocking labour (dependent variable) is a function of age, deadweight, type, renewal works (independent variables). Mathematically, the relationships mentioned above can be expressed in the equation form (Eqs. (6.2)– (6.8)).

$$D_{\text{LABOUR}} = a + b * S_{\text{A}} \tag{6.2}$$

$$D_{\text{LABOUR}} = a + b * S_{\text{D}} \tag{6.3}$$

$$D_{\text{LABOUR}} = a + b * (S_{\text{A}} * S_{\text{D}}) \tag{6.4}$$

$$D_{\text{LABOUR}} = a + b * S_{\text{T}} \tag{6.5}$$

$$D_{\text{LABOUR}} = a + b * R_{\text{hb}} \tag{6.6}$$

$$D_{\text{LABOUR}} = a + b * R_{\text{hp}} \tag{6.7}$$

$$D_{\text{LABOUR}} = a + b * R_{\text{hc}} \tag{6.8}$$

Since all the independent variables are linearly associated with the dependent variable, so it is expected a multiple linear regression model will be an excellent fit for the system. Accordingly, a multiple linear regression model is considered to establish the relationship between drydocking labour, age, deadweight, hull blasting and painting renewal area. The following functions (Eqs. (6.9), (6.10) and (6.11)) are chosen because the drydocking labour is a function of each independent variable as per primarily mentioned assumptions. Equation (6.9) is designed when drydocking labour can be estimated using age, deadweight, and type only. Equation (6.10) is designed when drydocking labour can be calculated using age, deadweight, and type together with the hull blasting and painting renewal area. Equation (6.11) is developed when drydocking labour can be estimated using age, deadweight, and type with hull coating (blasting and painting together) renewal area.

$$D_{\text{LABOUR}} = f(S_{\text{A}}, S_{\text{D}}, S_{\text{T}}) \tag{6.9}$$

$$D_{\text{LABOUR}} = f\left(S_{\text{A}}, S_{\text{D}}, S_{\text{T}}, R_{\text{hb}}, R_{\text{hp}}\right) \tag{6.10}$$

Table 6.3 Numerical values for types of ships for regression analysis

Types of ships	Numerical values for S_T for	
	Equation (6.9)	Equations (6.10) and (6.11)
Car carrier	1.0000	1.0000
Bulk carrier	1.1661	1.1661
General cargo carrier	1.2219	1.2219
Crude oil tanker	1.2812	1.2812
Chemical tanker	1.3979	1.3238
Container carrier	1.9055	1.9055
LPG carrier	2.0081	2.0844

$$D_{LABOUR} = f(S_A, S_D, S_T, R_{hc}) \qquad (6.11)$$

Appropriate numerical values for S_T are calculated and assigned for types of ships for regression analysis. Table 6.3 displays the numerical values assigned to the ship types for the functional equations (Eqs. (6.9)–(6.11)). It is used in regression analysis to form the regression Eqs. (6.12)–(6.14).

The final regression equations are formed following the regression analysis method described in Chap. 3 and using the observed data for D_{LABOUR}, S_A, S_D, S_T, R_{hb}, R_{hp} and R_{hc}. They are as follows,

$$D_{LABOUR} = -374.910 + 143.890 * S_A + 0.002 * S_D + 356.780 * S_T \quad (6.12)$$

$$D_{LABOUR} = -132.340 + 135.940 * S_A - 0.002 * S_D - 25.661 * S_T + \\ 0.056 * R_{hb} + 0.020 * R_{hp} \qquad (6.13)$$

$$D_{LABOUR} = -146.891 + 140.640 * S_A - 0.003 * S_D - 80.346 * S_T \\ + 0.026 * R_{hc} \qquad (6.14)$$

The vital statistical parameters regarding final regression equations (Eqs. (6.12), (6.13) and (6.14)) are given in Table 6.4. Significant improvement in coefficient of determination (R^2) is observed due to hull blasting, and painting renewal works Eqs. (6.13) and (6.14). The condition $f > f_{0.05}$ (calculated F statistic and tabulated F statistic, respectively) suggested rejecting the null hypothesis. It may be concluded that there is a significant amount of variation in their response (the dependent variable) due to the differences in independent variables in the postulated models.

Table 6.4 Summary of values of statistical parameters of final regression equations

Statistical parameters	Values of statistical parameters for		
	Equation (6.12)	Equation (6.13)	Equation (6.14)
Sample size (n)	50	48	48
No. of independent variable (k)	3	5	4
Significance level (α)	0.05	0.05	0.05
Standard deviation (s)	886	823	815
Coefficient of determination (R^2)	0.516	0.618	0.616
$f - f_\alpha$	13.55	11.12	14.64

6.9 Validation

The validation technique is applied to Eqs. (6.12)–(6.14). The summary of the validation results is presented in Table 6.5. The table shows the outline of the variation of model values from the actual values. It is important to note that Eq. (6.12) considers only age, deadweight, and type of ship irrespective of repairing activities. Equation (6.13) assumes age, deadweight, type together with hull coating renewal area. The improvement in validation results of Eq. (6.13) over (6.12) is significant in all parameters. In contrast, Eq. (6.14) did not show any improvement compared to Eq. (6.13). Practically, Eqs. (6.13) and (6.14) yield similar validation results because the combined effects of blasting and painting are identical to coating.

The probable reasons for deviation of model values from actual values explained in previous sections are equally applicable to drydocking labour, too. Only nature and magnitudes are different. Figure 6.24 demonstrates the relationship of deviation (%) in drydocking labour against equivalent variation in independent variables.

Validation is also applied for drydocking labour-age relationship under a linear form of the equation for all types together (Fig. 6.1) and different kinds of ships such as crude oil tankers (Fig. 6.5), container carriers (Fig. 6.6), chemical tankers (Fig. 6.7), bulk carriers (Fig. 6.8) and liquified petroleum gas carriers (Fig. 6.9). The results are presented in Table 6.6. It displays that validation result under individual

Table 6.5 Summary of validation results of final regression equations

Items	Equation (6.12)	Equation (6.13)	Equation (6.14)
Positive error (%)	73	213	212
Negative error (%)	−149	−62	−64
Range of error (%)	222	275	276
Mean error (%)	−19	16	16
Variance	2994	3492	3498
Standard deviation	55	59	59

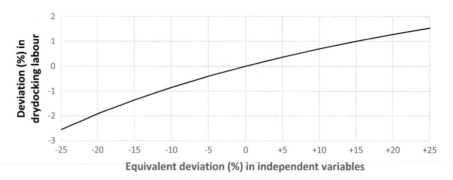

Fig. 6.24 Deviation in drydocking labour versus equivalent deviation in independent variables

Table 6.6 Summary of validation results of drydocking labour-age relationship

Type of ships →	All types	COT	CC	ChT	BC	LPGC
Item ↓/Fig. No. →	6.1	6.5	6.6	6.7	6.8	6.9
Positive error (%)	185	134	33	158	85	94
Negative error (%)	−65	−53	−60	−71	−39	−42
Range of error (%)	250	187	93	229	124	137
Mean error (%)	13	17	−7	10	14	13
Variance	2971	3412	876	6703	2065	1993
Standard deviation	55	58	30	82	45	45

type is better than all kinds, including the range of error, mean error, and standard deviation of error except chemical tanker.

The same technique is applied to the drydocking labour-deadweight relationship under a linear form of equation (Fig. 6.2) and for different types of ships such as crude oil tankers (Fig. 6.10), container carriers (Fig. 6.11), chemical tankers (Fig. 6.12), bulk carriers (Fig. 6.13) and liquified petroleum gas carriers (Fig. 6.14). The results are presented in Table 6.7. It displays that validation result under individual type

Table 6.7 Summary of validation results of drydocking labour-deadweight relationship

Type of ships →	All types	COT	CC	ChT	BC	LPGC
Item ↓/Fig. No. →	6.2	6.10	6.11	6.12	6.13	6.14
Positive error (%)	247	−10	−3209	−971	−322	−2329
Negative error (%)	−139	−159	−3473	−1325	−438	−2435
Range of error (%)	386	169	6682	2295	760	4764
Mean error (%)	−44	−85	−3386	−1212	−382	−2384
Variance	5338	1612	7047	10,527	1894	1677
Standard deviation	73	40	84	103	44	41

does not improve, even worse than all types except bulk carriers. This situation demonstrates that a ship's deadweight does not dictate the drydocking labour under individual type.

Finally, the proposed mathematical models may estimate repairing labour for a ship undergoing a routine maintenance program. Using the above model as a guide, shipyards may evaluate the expected drydocking labour against age, deadweight and anticipated scope of hull blasting area and painting area. While using the model to estimate the expected drydocking labour, one may be aware of the error level in predicting drydocking labour. The model will reasonably estimate drydocking labour if the desired independent variables are close to the mean value. However, for low and high values, the assessment will be low and high accordingly (Fig. 6.24). Also, one may consider allowing some allowance on top of the model value to accommodate the various variation of drydocking labour explained earlier.

6.10 General Conclusions

The most influential relationships of the dependent variable (drydocking labour) and various independent variables (age, deadweight, type, hull blasting and painting renewal area) are identified through analysis. Three multiple linear regression equations are developed of different combinations of independent variables. The validation technique is applied to appropriate equations to demonstrate the effectiveness. New Figures are developed for users to estimate drydocking labour for planning and budgetary purpose. Table 6.8 displays the correlation coefficients for drydocking labour-age and drydocking labour-deadweight under linear relationship for types. It provides an instant idea of how a ship's age and deadweight may influence the expected drydocking labour.

Figure 6.25 is constructed using the linear relationship of Fig. 6.1. It demonstrates the estimated drydocking labour against age irrespective of deadweight, type, and other activities.

Table 6.8 Summary of correlation coefficients for drydocking labour-age and deadweight in a linear relationship for types

Types of ship	Correlation coefficient (r^2)	
	$D_{LABOUR} = f(S_A)$	$D_{LABOUR} = f(S_D)$
All types	0.830	0.084
Crude oil tanker	0.217	0.929
Container carrier	0.743	0.183
Chemical tanker	0.817	0.361
Bulk carrier	0.415	0.700
LPG carrier	0.727	0.277

Fig. 6.25 Estimated drydocking labour versus age

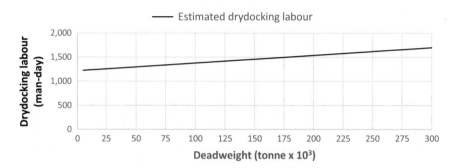

Fig. 6.26 Estimated drydocking labour versus deadweight

Figure 6.26 is constructed using the linear relationship of Fig. 6.2. It demonstrates the estimated drydocking labour against deadweight irrespective of age, type, and other activities.

Figure 6.27 is constructed using the linear relationship of Fig. 6.3. It demonstrates the estimated drydocking labour against (age * deadweight) irrespective of type and other activities.

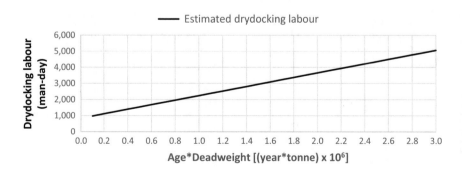

Fig. 6.27 Estimated drydocking labour versus (age * deadweight)

Figure 6.28 is constructed using the linear relationship of Figs. 6.5, 6.6, 6.7, 6.8 and 6.9 for crude oil tankers, container carriers, chemical tankers, bulk carriers and liquified petroleum gas carriers, respectively. It demonstrates the estimated drydocking labour against age for mentioned types irrespective of deadweight and other activities.

Figure 6.29 is constructed using the linear relationship of Figs. 6.10, 6.11, 6.12, 6.13 and 6.14 for crude oil tankers, container carriers, chemical tankers, bulk carriers and liquified petroleum gas carriers, respectively. It demonstrates the estimated drydocking labour against deadweight for mentioned types irrespective of age and other activities.

Estimation of drydocking labour will vary with the available independent variables. At the preliminary stage, many variables are not available. One may like to

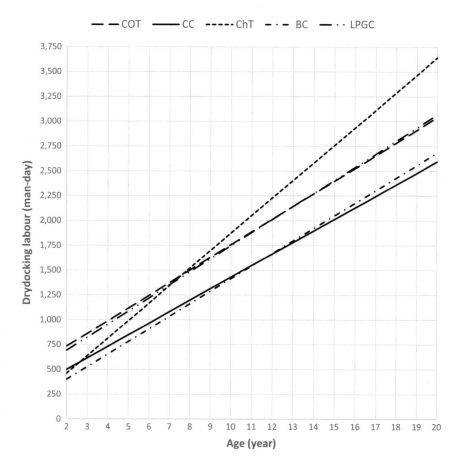

Fig. 6.28 Estimated drydocking labour versus age for crude oil tankers, container carriers, chemical tankers, bulk carriers and liquified petroleum gas carriers

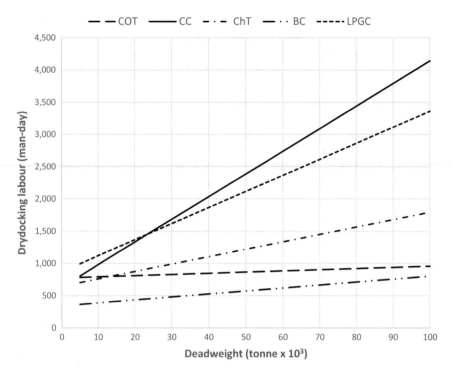

Fig. 6.29 Estimated drydocking labour versus deadweight for crude oil tankers, container carriers, chemical tankers, bulk carriers and liquified petroleum gas carriers

follow the below options to estimate drydocking labour under various conditions of independent variables.

Option—I
Use age and estimate the drydocking labour irrespective of deadweight, type, and other activities with the help of Fig. 6.25.

Option—II
Use deadweight and estimate the drydocking labour irrespective of age, type, and other activities with the help of Fig. 6.26.

Option—III
Use (age * deadweight) and estimate the drydocking labour irrespective of type and other activities with the help of Fig. 6.27.

Option—IV
Use age and type and estimate the drydocking labour for the corresponding type, irrespective of deadweight and other activities, with the help of Fig. 6.28 as appropriate.

Option—V

Use deadweight and type and estimate the drydocking labour for the corresponding type, irrespective of age and other activities, with the help of Fig. 6.29 as appropriate.

Option—VI

Use age, deadweight and type and estimate the drydocking labour for the corresponding type, irrespective of other activities, with regression Eq. (6.12).

Option—VII

Use age, deadweight, type, hull blasting and painting renewal area and estimate the drydocking labour, irrespective of other activities, with regression Eq. (6.13).

Option—VIII

Use age, deadweight, type, hull coating renewal area and estimate the drydocking labour with regression Eq. (6.14).

Practically, Eqs. (6.13) and (6.14) will yield the same results.

References

1. Dev, A.K., Saha, M.: Dry-docking time and labour, Trans RINA, Vol. 164, Part 4. Int. J. Maritime Eng. **2018**, 337–380 (2018)
2. Dev, A.K., Saha, M.: Modeling and analysis of ship repairing labor. J. Ship Prod. Des. **32**(4), 258–271 (2016)

Chapter 7
Hull Coating Renewal Area

7.1 Introduction

In general, the hull coating of a ship refers to the hull protection of a ship using paints applied on the hull surface during new building time. It is required to re-apply after the paint life is over or paint damage within the paint life. This process is repeated during the service/operational life of a ship. During new building time, the entire hull area is involved regarding blasting and painting, no spot blasting and touch up coat, but in case of repairing, part or entire hull area is involved for blasting followed by touch-up coat(s) and full coat(s). In this Chapter, the focus will be on the repairing of hull coating only from various viewpoints.

The hull coating process is divided into two parts, (i) Surface preparation using blasting technique (grit blasting, hydro-blasting) to achieve a required standard and (ii) Application of paint (anti-corrosive, anti-fouling) to cover the blasted area with the required number of layers (number of coats) to develop a required film thickness as per required paint life.

Basic hull coating repairing process/system in routine maintenance of a ship can be explained as follows: upon inspection and confirmation of the scope of hull-coating repairing works by a representative of paint manufacturer and approved by the shipowner's representative (generally, the attending superintendent), the shipyard starts the activity with blasting, such as spot blasting (to remove the paints from the rusty spot, paint damaged area, etc.) or complete blasting (to remove the paint from the entire area) to get a bare metal surface. Blasting quality may vary from SA 1.0 to SA 2.5 depending on the painting condition at the time of inspection and decided by the paint manufacturer. After completing blasting, the paint application starts as per paint schemes such as single or multiple coats of anti-corrosive paint to the spot blasted area and single or multiple coats of anti-corrosive and anti-fouling paint to the entire hull area.

Figure 7.1 shows a typical midship section with conventional load lines and hull area demarcations. The topside area is the ship side from the deep load line to the main

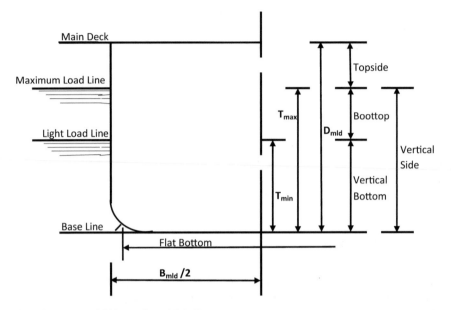

Fig. 7.1 Typical midship section with hull area

deck and is consistently above water. No anti-fouling paint but only anti-corrosive paint is required to apply in this area. Boot top area is the area at the ship side between light load line to deep load line and constantly subjected to be in and out of water alternately. The topcoat in this area is of different quality to meet the performance requirements. It is worth mentioning that the boot top area is not always demarked separately for some ships but included in the vertical bottom area under the anti-fouling paint. The vertical bottom area is the area at the ship side between baseline to light load line and permanently underwater. The topcoat in this area is anti-fouling paint. Vertical bottom area together with boot top area is often called vertical side area and covered with anti-fouling paint. The flat bottom area is the baseline level (generally from port bilge keel to starboard bilge keel). Need not to mention that it is always underwater and covered with anti-fouling paint. The repairing of hull-coating refers to the blasting of the designated area and an anti-corrosive and anti-fouling painting system.

Hull-coating repair is a part of the routine maintenance work of a ship during its working life. Due to the location of the repairing works, this can only be carried out in a dock (graving dock, floating dock, and slipway). For routine maintenance drydocking schedule, external hull-coating repair (blasting and painting) is critical in deciding a ship's duration of stay in the drydock. The scope of hull-coating repairing works usually dictates the drydocking time. Due to the cost factor, more drydocking time means more cost, and as such, both the shipowner and shipyard would like to minimise the drydocking time.

Due to the location of the hull coating works, it is simply impossible to know the accurate scope of coating renewal works. Moreover, there are no guidelines available that could be useful for the ship owners and ship managers for a reliable estimation of the scope of hull-coating repairing works before drydocking the ship. The probable reason seems to be the scarcity and confidentiality of such commercially sensitive information and data. Typically, details of coating repair such as blasting quality and quantity, types of paint, the number of coats, etc., are not disclosed to other than the concerned parties. As such, there is little opportunity for information sharing of this valuable ship maintenance item.

No documented information about the estimation of hull-coating repairing area of ships is available regarding their age, deadweight, and type. Some works, not exactly but close to the issue, were done from different viewpoints. Broderick [1] explored the link between structural complexity in water ballast tanks and coating performance in the context of the introduction of the IMO Performance Standard for Protective Coatings (PSPC) for dedicated water ballast tanks (WBT), intending to propose how future ship structural design may be improved to enhance coating performance. This paper highlighted the classification of complexity of coating surface described by Jotun [2], where the flats of the ship's hull are classed as low, the cargo holds of a bulk carrier are given a medium rating, and WBTs are classed as very high. It also suggested only a relatively small feasible region in the design space within which alternative stiffener types and scantlings could be proposed to seek such benefit on the coating performance. Dev and Saha [3] reviewed the hull coating renewal area during routine maintenance. This article attempts to demonstrate the trends of hull coating renewal in ship repairing concerning age, deadweight, and type of ships. It suggests that hull coating repairing works are a function of ships' age, deadweight, type, and principal dimensions but at different degrees of responses. It also reveals some fundamental basis for estimating average hull blasting and painting areas for various ships' age, deadweight, type and principal dimensions. All independent variables are mostly linearly associated with the dependent variable. Garbatov et al. [4] evaluated corrosion development and wastage of deck plates of ballast water tanks (WBT) and cargo oil tanks (COT) for tankers. They developed a non-linear time-dependent corrosion wastage model for the deck plate. The model can describe an initial period without corrosion due to the presence of a corrosion protection system (tank coating), a transition period with a non-linear increase in wastage up to a steady-state of long-run corrosion wastage that leads to plate replacement. The model also predicts that the time periods without corrosion for the WBT deck plate and the COT deck plate, which corresponds to the start of failure of corrosion protection coating, are 10.54 years and 11.49 years, respectively, and the transition periods for the same are 11.14 years and 11.23 years respectively. Hiromi et al. [5] researched a reduction in wall thickness of various shipboard piping systems resulting from flow-accelerated corrosion (FAC) under different flow conditions and pipe geometry. They also proposed using Kastner's experimental formula to estimate the reduction in wall thickness of pipelines onboard. Nakai et al. [6] highlighted the corroded condition of webs of cargo hold frames of a bulk carrier and investigated the effect of corrosion pitting and its contribution to ultimate strength, particularly the webs of hold frames of a bulk carrier that carries iron ore and coal. They developed a visual assessment method of corroded conditions and various parameters required to calculate the ultimate strength. They also prepared and proposed a method to estimate the equivalent thickness loss of web plate using the depth of pitting and the residual strength of members with pitting corrosion. Their predicted result of tensile strength strongly corresponds with the experimental result. O'Donnell [7] provides an overview of the corrosion fatigue of the most commonly used carbon and low-alloy steels and stainless steel and its developments and future needs considering environmentally assisted cracking (ESC). It presented compilations and analysis of available databases and suggested environmental fatigue curves for carbon and alloy steels and austenitic stainless steels. Paik et al. [8] investigated the effect of corrosion on the ultimate strength of a structural member of ships' structure. They developed and demonstrated a procedure for assessing ship hull girder ultimate strength reliability considering the degradation of primary members due to general corrosion. A probabilistic model for ultimate hull girder strength is established by an analytical formula that considers corrosion-related time-dependent strength degradation in various failure modes. The variability in strength, corrosion rates and loads are accounted for in the second-order reliability method (SORM) based on the time-dependent reliability index calculations. The procedure developed is illustrated by application to both tankers and bulk carriers. For a given set of renewal criteria, apart from hull girder section modulus trends, ultimate strength, and the reliability index as a function of ship age, the probability of steel renewal due to corrosion is also predicted. Yamamoto et al. [9] demonstrated the corrosion phenomenon of a ship's hull consisting of three sequential processes such as (i) degradation of paint coatings, (ii) generation of pitting points and (iii) progress of pitting point. They described each process by introducing a probabilistic corrosion model. This probabilistic corrosion model can be developed by analysing existing data collected from plate thickness measurements. By comparing the results of estimations by the identified probabilistic models and actual measurement data, the practical usefulness of the proposed procedure is proved.

This Chapter will discuss how the age, deadweight, and type of a ship influence the hull coating renewal area (blasting area and painting area). First, there will be some assumptions, and subsequently, those will be verified by analysing the hull coating renewal area against the respective variable of sample data. In this analysis, the hull coating renewal area (blasting and painting independently) is considered the dependent variable, and others thought independent variables. Finally, a mathematical model will be presented to calculate the hull coating renewal area against a set of variables.

The general assumption is that hull blasting renewal area, hull painting renewal area, and hull coating renewal area is a function of age, deadweight, type, (age * deadweight), and linearly associated. In other words, age, deadweight, type and (age * deadweight) influence hull blasting, painting, and coating separately with a linear relationship. Therefore, in the following sections, each of them will be discussed.

7.2 Hull Blasting (R_{hb})

Primary hull blasting is meant to prepare the surface for the application of paint. Blasting quality may vary from standard SA 1.0 to SA 2.5. Generally, SA 2.5 is applied to achieve a bare metal surface. Hydro-blasting at various pressure is also used to prepare the surface. Hydro-blasting at 45,000 psi (pounds per square inch) can produce a surface equivalent to SA 2.5 (Authors have experience using this method). By nature, hydro-blasting is a slow and risky process. In present days, remote control hydro-blasting machines (Octopus type) are used with low risk. As a practice, the paint manufacturer decides the quality of surface preparation to be achieved. Generally, to express the importance of quality of surface preparation for coating, there is a saying, "A bad paint on a good surface is much better than a good paint on a bad surface", and that is why surface preparation is the most critical and essential part of the coating process.

As per general assumption, the hull blasting renewal area is a function of age, deadweight, type, (age * deadweight), and linearly associated. In other words, age, deadweight, type and (age * deadweight) influence the blasting renewal area with a linear relationship. Therefore, in the following sections, each of them will be discussed.

7.2.1 Hull Blasting Renewal Area Versus Age (R_{hb} vs S_A)

The hull blasting renewal area is a function of age regardless of deadweight and type as per primary assumption.

Initial investigation of the hull blasting renewal area and age is demonstrated in Figs. 7.2 and 7.3, showing the hull blasting renewal area (absolute value and as a percentage of total hull area) against age, respectively, irrespective of deadweight and type. Figures show a positive relationship in both cases. It is expected for a simple reason. As a ship becomes older, the quality of steel surface deteriorates, from various viewpoints, such as chemical properties (metallurgical), mechanical properties (structural strength), physical properties (surface roughness, bonding to coating, etc.) due to natural causes. Older ships will likely have more blasting and painting works than new ships. The linear equations, $R_{hb} = 731.040 + 245.830 * S_A$ and $R_{hb}(\%) = 0.020 + 1.608 * S_A$, provide the best goodness of fit to the sample data with correlation coefficients of 0.488 and 0.545 respectively.

Therefore, the assumption made is valid. However, more clearly, older ships will demand more blasting renewal works than newer ships.

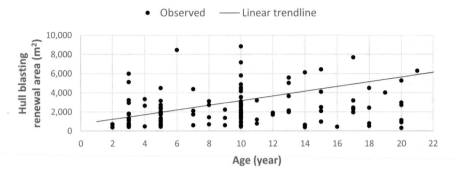

Fig. 7.2 Hull blasting renewal area versus age

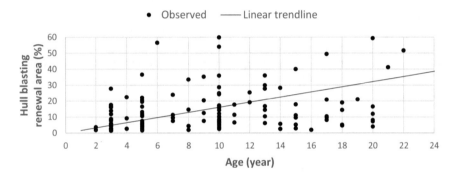

Fig. 7.3 Hull blasting renewal area versus age

7.2.2   Hull Blasting Renewal Area Versus Deadweight (R_{hb} vs S_D)

It is assumed that the deadweight will influence the hull blasting renewal area irrespective of age and type.

Examination of the hull blasting renewal area and deadweight is displayed in Figs. 7.4 and 7.5, showing the hull blasting renewal area (absolute value and as a percentage of total hull area) against deadweight, respectively, irrespective of age and type. Figures show a positive relationship in both cases. It is very likely because higher deadweight means bigger size resulting larger external surface areas. Thus, logically the bigger ships are likely to have more hull blasting renewal areas. Hence, the deadweight will influence the hull blasting renewal area. Therefore, a ship's hull blasting renewal area is a function of deadweight irrespective of age and type and linearly associated. The linear equations, $R_{hb} = 1697.600 + 7.143 * \left(S_D/10^3 \right)$ and $R_{hb}(\%) = 12.542 + 0.006 * \left(S_D/10^3 \right)$, provide the best goodness of fit to the sample data with correlation coefficients of 0.312 and 0.045 respectively.

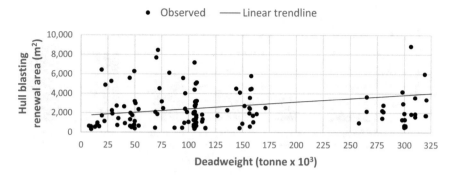

Fig. 7.4 Hull blasting renewal area versus deadweight

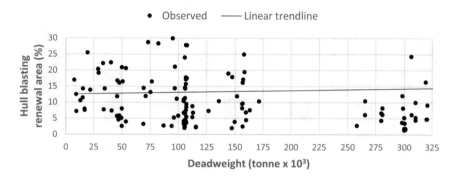

Fig. 7.5 Hull blasting renewal area versus deadweight

Therefore, the assumption made is valid. So, bigger ships will have more blasting renewal works than smaller ships.

7.2.3 Hull Blasting Renewal Area Versus Type (R_{hb} vs S_T)

As per general assumptions, the type of a ship influences hull blasting renewal area. It means that the different types of ships with the same age and deadweight will have a different hull-coating renewal scope.

Analysis of hull blasting renewal area against type is depicted in Figs. 7.6 and 7.7, showing the hull blasting renewal area (absolute value and as a percentage of total hull area) against type, respectively, irrespective of age and deadweight. Again, figures show a strong relationship. It is expected due to the inherent differences in the kinds of ships.

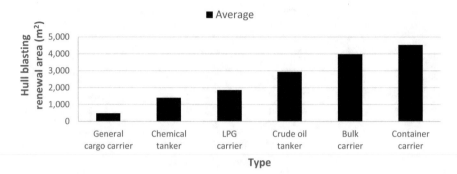

Fig. 7.6 Average hull blasting renewal area versus type

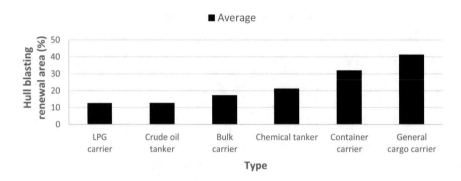

Fig. 7.7 Average hull blasting renewal area versus type

Therefore, the assumption made is valid and different types of ships will demand different scopes of hull blasting renewal work even if they are of the same age and size.

7.2.4   Hull Blasting Renewal Area Versus (age * deadweight) [R_{hb} vs (S_A*S_D)]

The general assumption suggests that the variable (age * deadweight) will influence the hull blasting renewal area with a linear association.

A study of the hull blasting renewal area and (age * deadweight) is presented in Figs. 7.8 and 7.9 showing the hull blasting renewal area (absolute value and as a percentage of total hull area) against (age * deadweight), respectively, irrespective of type. Figures show a positive relationship for both cases. It is logical to state that the higher value of (age * deadweight) demands older and bigger ships leading to a higher hull blasting renewal area. The linear equations, $R_{hb} = 153.080 +$

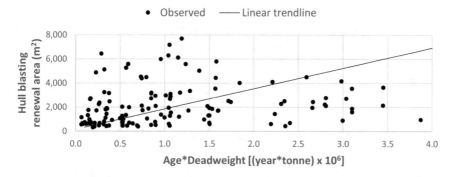

Fig. 7.8 Hull blasting renewal area versus (age * deadweight)

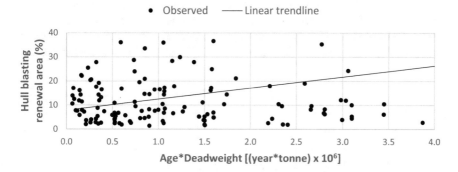

Fig. 7.9 Hull blasting renewal area versus (age * deadweight)

$1694.100*\left[(S_A * S_D)/10^6\right]$ and $R_{hb}(\%) = 7.868+4.613*\left[(S_A * S_D)/10^6\right]$, provide the best goodness of fit to sample data with correlation coefficients of 0.451 and 0.389 respectively.

Therefore, the assumption made is valid. So, bigger and older ships will have more blasting renewal works than smaller and newer ships.

7.3 Hull Painting (R_{hp})

Hull painting is the finishing stage of the hull coating renewal works. Generally, single coat or multiple coats of anti-corrosive and anti-fouling paints are applied to the blasted area and total area as per the painting scheme. For example, at the topside area, numerous coats of anti-corrosive paints are applied only. Then, multiple coats of anti-corrosive paints are applied in other areas, followed by numerous multiple coats of anti-fouling paints. The number of coats (touch-up and complete) is strictly decided by the paint manufacturer and the shipowner's representative.

As per the general assumption, the hull painting renewal area is a function of age, deadweight, type, (age * deadweight), and linearly associated. In other words, age, deadweight, type and (age * deadweight) each influence hull painting renewal area with a linear association. Therefore, in the following sections, each of them will be discussed.

7.3.1 Hull Painting Renewal Area Versus Age (R_{hp} vs S_A)

As per general assumption, the age of a ship influences hull painting renewal area irrespective of deadweight and type. Therefore, the hull painting renewal area is a function of age regardless of deadweight, type, and linearly associated.

Initial investigation of hull painting renewal area versus age is demonstrated in Figs. 7.10 and 7.11 showing the hull painting renewal area (absolute value and as a percentage of total hull area) against age, respectively, irrespective of deadweight and type. Again, the figures show a positive relationship. This is expected because,

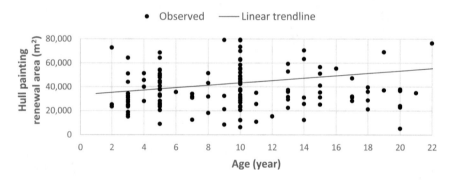

Fig. 7.10 Hull painting renewal area versus age

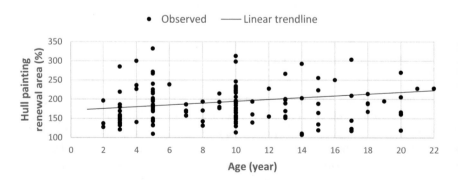

Fig. 7.11 Hull painting renewal area versus age

unlike the hull blasting area, the painting area does not depend on how old the ship is but rather considerably on how big the ship is. This is demonstrated by the lower value of regression coefficient for hull painting compared to hull blasting. The linear equations, $R_{hp} = 33305.000 + 1004.200 * S_A$ and $R_{hp}(\%) = 171.110 + 2.349 * S_A$, provide the best goodness of fit to sample data with correlation coefficients of 0.284 and 0.178 respectively.

Therefore, the assumption is valid. However, more clearly, older ships are likely but not necessarily to have more painting renewal works than newer ships.

7.3.2 Hull Painting Renewal Area Versus Deadweight (R_{hp} vs S_D)

The hull painting renewal area is assumed to be a function of deadweight irrespective of age and type and linearly associated.

Examination of hull painting renewal area and deadweight is presented in Figs. 7.12 and 7.13 showing the hull painting renewal area (absolute value and as a percentage of total hull area) against deadweight, respectively, irrespective of age and type. Figures show a positive relationship for the total area but negative for the percentage area. This is expected because the higher deadweight demands bigger size and more surface area and, eventually, higher hull painting renewal area. Therefore, mathematically, a negative slope means the rate of change in hull painting renewal area concerning deadweight is lower than that of deadweight. The linear equations, $R_{hp} = 20665.000 + 134.360 * (S_D/10^3)$ and $R_{hp}(\%) = 217.550 - 0.165 * (S_D/10^3)$, provide the best goodness of fit relationship with correlation coefficients of 0.869 and 0.177 respectively.

Therefore, the assumption made is valid. As such, bigger ships are most likely to have more painting renewal works than smaller ships.

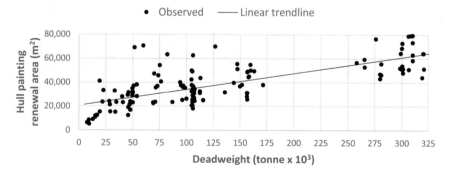

Fig. 7.12 Hull painting renewal area versus deadweight

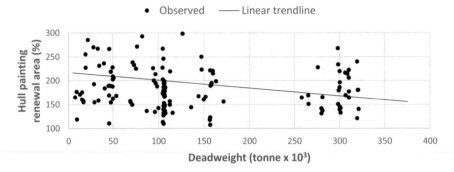

Fig. 7.13   Hull painting renewal area versus deadweight

7.3.3   Hull Painting Renewal Area Versus Type (R_{hp} vs S_T)

As per assumption, the type of a ship impacts the hull painting renewal area irrespective of age and deadweight. Therefore, a ship's hull painting renewal area is a function of type regardless of age and deadweight and linearly associated.

Analysis of hull painting renewal area against type is displayed in Figs. 7.14 and 7.15 showing the hull blasting renewal area (absolute value and as a percentage of total hull area) against type, respectively, irrespective of age and deadweight. Again, figures show a strong relationship. It is expected due to the inherent differences in the kinds of ships.

Therefore, the assumption made is valid and different types of ships will demand different scopes of hull painting renewal work even if they are of the same age and size.

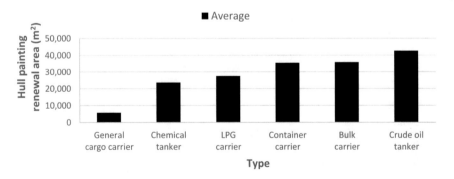

Fig. 7.14   Average hull painting renewal area versus type

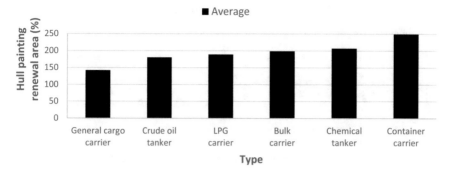

Fig. 7.15 Average hull painting renewal area versus type

7.3.4   Hull Painting Renewal Area Versus (age * deadweight) [R_{hp} vs (S_A*S_D)]

Initial assumptions suggest that the hull painting renewal area is a linearly associated function of (age * deadweight).

Study of hull painting renewal area and (age * deadweight) is depicted in Figs. 7.16 and 7.17 showing the hull painting renewal area (absolute value and as a percentage of total hull area) against (age * deadweight) respectively, irrespective of type. Figures show a positive relationship for both. Mathematically, it is true that the higher value of (age * deadweight) demands older and bigger ships leading to a higher hull painting renewal area. The linear equations, $R_{hp} = 28717.000 + 7970.500 * \left[(S_A * S_D)/10^6\right]$ and $R_{hp}(\%) = 189.510 + 5.215 * \left[(S_A * S_D)/10^6\right]$, provide the best goodness of fit relationship with correlation coefficients of 0.651 and 0.558 respectively.

Therefore, the assumption made is valid, so bigger and older ships will have more painting renewal works than smaller and newer ships.

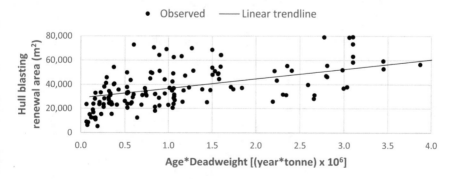

Fig. 7.16 Hull painting renewal area versus (age * deadweight)

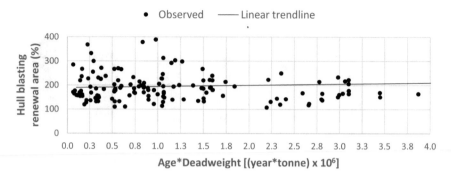

Fig. 7.17 Hull painting renewal area versus (age * deadweight)

7.4 Hull Coating (R_{hc})

As hull coating works refer to the hull blasting works and hull painting works together, hull coating works are also a function of age, deadweight, type and (age * deadweight) and are linearly associated. Therefore, this sub-section will focus on how the hull coating works are affected by age, deadweight, type and (age * deadweight). In the following sub-sections, each of them will be discussed.

7.4.1 Hull Coating Renewal Area Versus Age (R_{hc} vs S_A)

The hull coating renewal works are a function of age irrespective of deadweight and type and linearly associated.

Initial investigation of hull coating renewal area versus age is demonstrated in Figs. 7.18 and 7.19 showing the hull coating renewal area (absolute value and as a percentage of total hull area) against age, respectively, irrespective of deadweight

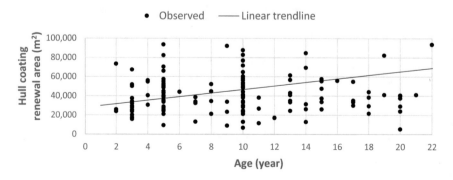

Fig. 7.18 Hull coating renewal area versus age

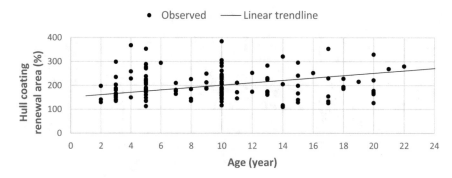

Fig. 7.19 Hull coating renewal area versus age

and type. Figures show a positive relationship. It is very much likely because the hull coating renewal area includes hull blasting and painting renewal area. Therefore, older ships will see a higher hull coating renewal area. The linear equations, $R_{hc} = 28020.000 + 1859.800 * S_A$ and $R_{hc}(\%) = 151.650 + 4.981 * S_A$, provide the best goodness of fit to sample data with correlation coefficients of 0.550 and 0.459 respectively.

Therefore, the assumption made is valid. However, more clearly, older ships will have more hull coating renewal works than newer ships.

7.4.2   Hull Coating Renewal Area Versus Deadweight (R_{hc} vs S_D)

The primary assumption is that the hull coating renewal works is a function of deadweight irrespective of age and type and linearly associated.

Initial analysis of hull coating renewal area and deadweight is displayed in Figs. 7.20 and 7.21 showing the hull coating renewal area (absolute value and as a percentage of total hull area) against deadweight, respectively, irrespective of age and type. Figures show a strong relationship. This is expected because blasting and painting are impacted by deadweight positively and linearly. The linear equations, $R_{hc} = 24004.000 + 134.030 * (S_D/10^3)$ and $R_{hp}(\%) = 237.810 - 0.204 * (S_D/10^3)$, provide the best goodness of fit to sample data with correlation coefficients of 0.839 and 0.207 respectively.

Therefore, the assumption made is valid. More specifically, bigger ships are expected to have more coating renewal works than smaller ships.

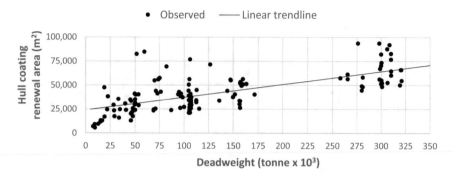

Fig. 7.20 Hull coating renewal area versus deadweight

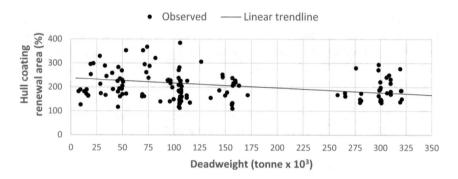

Fig. 7.21 Hull coating renewal area versus deadweight

7.4.3 *Hull Coating Renewal Area Versus Type (R_{hc} vs S_T)*

As per assumption, the hull coating renewal works is a function of type irrespective of age and deadweight and linearly associated.

Examination of hull coating renewal area against type is depicted in Figs. 7.22 and 7.23 showing the hull coating renewal area (absolute value and as a percentage of total hull area) against type, respectively, irrespective of age and deadweight. Figures show a strong relationship. It is very much standard due to the inherent differences in the kinds of ships.

Therefore, the assumption made is valid and different types of ships will demand different scopes of hull coating renewal work even if they are of the same age and size.

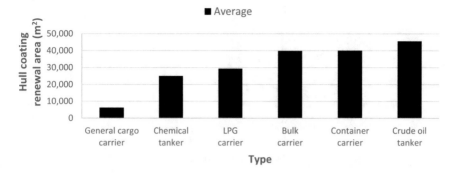

Fig. 7.22 Average hull coating renewal area versus type

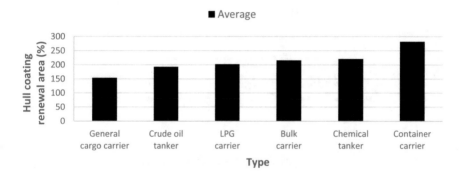

Fig. 7.23 Average hull coating renewal area versus type

7.4.4 *Hull Coating Renewal Area Versus (age * deadweight)* *[R_{hc} vs (S_A * S_D)]*

The assumption is that the hull coating renewal works is a function of (age * deadweight) irrespective of type with a linear association.

Study of hull coating renewal area and (age * deadweight) is highlighted in Figs. 7.24 and 7.25, showing the hull coating renewal area (absolute value and as a percentage of total hull area) against (age * deadweight), respectively, irrespective of type. Figures show a positive relationship for the total area but negative for the percentage area. Mathematically, it is obvious because the higher value of (age * deadweight) demands older and bigger ships leading to a higher hull painting renewal area. The linear equations, $R_{hc} = 32306.000 + 9092.900 * [(S_A * S_D)/10^6]$ and R_{hp} (%) $= 198.280 + 9.229 * [(S_A * S_D)/10^6]$ provide the best goodness of fit to sample data with correlation coefficients of 0.746 and 0.433 respectively.

Therefore, the assumption made is valid and bigger, and older ships will have more coating renewal works than smaller and newer ships.

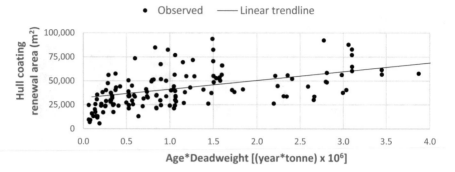

Fig. 7.24 Hull coating renewal area versus (age * deadweight)

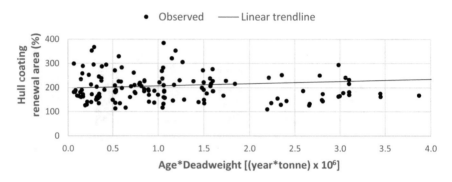

Fig. 7.25 Hull coating renewal area versus (age * deadweight)

7.5 Hull Coating Versus Dimensions

7.5.1 Introduction

There is no doubt that the quantity of blasting and painting repairing works is significantly influenced by a ship's physical size, irrespective of its design parameters. Physically bigger ships will require more blasting and painting repairing works compared to smaller ships. The size of a ship can be defined in various ways using different parameters, such as displacement, deadweight, gross tonnage, and principal dimensions. In this sub-section, the physical size of a ship refers to the physical dimensions of a ship, such as length, breadth, and depth. Without any assumptions, it can be stated firmly that blasting, painting, and coating (blasting and painting jointly) repairing works are entirely a function of ships' principal dimensions and hull total area, respectively. Accordingly, blasting, painting, and coating repairing areas of the hull and other hull locations are analysed as a function of respective ships' principal dimensions and total area. In other words, blasting, painting, and coating repairing area of the hull and its hull locations area are considered dependent variables and corresponding ships' principal dimensions with appropriate combinations and particular hull location actual area are considered independent variables. Various functional relationships are drawn using these multiple sets of dependent variables and independent variables.

The functional relationship of different hull locations will be other with varying combinations of principal dimensions of a ship. The functional equations are shown below (a sample) illustrate a mathematical relationship between hull blasting and hull painting renewal area (the dependent variable) and corresponding ship's dimensions and hull location actual area (the independent variable), respectively. It is essential to mention and remember that the functional relationship mentioned above, and equations are shown below are not based on hypothesis or assumption but mathematical reality irrespective of age, deadweight, and type.

$$R_{hb} = f[L_{OA} * (B_{mld} + 2 * D_{mld})]$$
$$R_{hb} = f(A_{HT})$$
$$R_{VBB} = f(L_{OA} * 2T_{min})$$
$$R_{VBB} = f(A_{VB})$$
$$R_{hp} = f[L_{OA} * (B_{mld} + 2 * D_{mld})]$$
$$R_{hp} = f(A_{HT})$$
$$R_{VBP} = f(L_{OA} * 2T_{min})$$
$$R_{VBP} = f(A_{VB})$$

7.5.2 Methodology

Based on the boundary condition applied to the functional relationship between blasting, painting, and coating repairing works, ships' dimensions and individual hull locations area, the equation must satisfy the requirement for a zero-hull area or a zero-hull location area, blasting, painting, and coating area must be zero. Mathematically, it refers to the fact that the line of the equation must pass through the origin (0, 0). Therefore, the required equation is considered to be $Y = m*X$ form, Where Y: quantity of blasting or painting or coating repairing area, m^2, X: corresponding ship's dimensions in appropriate combinations or actual individual area, m^2 and m: a constant slope.

7.5.3 Hull Blasting

Initial investigation on pairs of variables of interest related to hull blasting renewal area is presented in Figs. 7.26, 7.27, 7.28, 7.29, 7.30, 7.31, 7.32, 7.33, 7.34, 7.35, 7.36

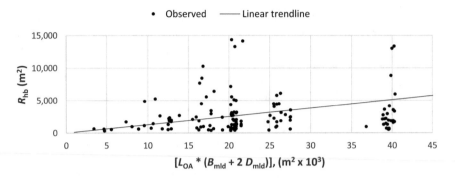

Fig. 7.26   Total hull blasting renewal area versus ship's dimensions

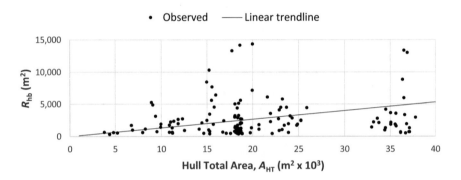

Fig. 7.27   Total hull blasting renewal area versus hull total area

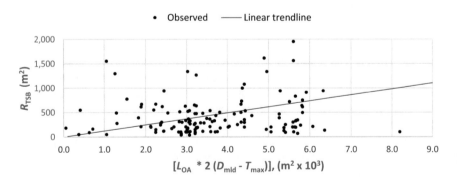

Fig. 7.28   Topside blasting renewal area versus ship's dimensions

and 7.37. It depicts the behaviour of the hull blasting renewal area against the ship's dimensions and particular hull location area, respectively. All these figures reveal that the blasting quantity also increases linearly with the increase of ships' dimensions. Thus, all these figures support the assumption made earlier that the size (length,

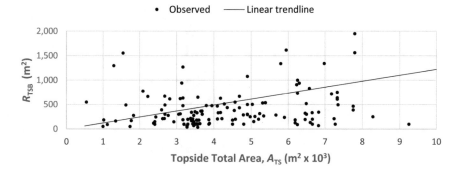

Fig. 7.29 Topside blasting renewal area versus topside total area

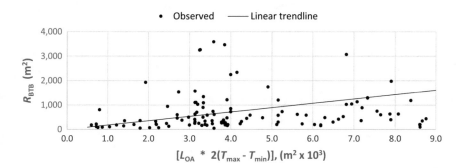

Fig. 7.30 Boottop blasting renewal area versus ship's dimensions

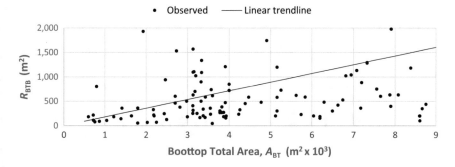

Fig. 7.31 Boottop blasting renewal area versus boottop total area

breadth, and depth) of a ship positively impacts hull blasting renewal area and is linearly associated. A summary of linear trendline equations and the corresponding correlation coefficients of relationships between hull blasting renewal area and ships' dimensions and actual areas is highlighted in Table 7.1.

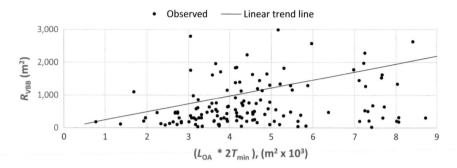

Fig. 7.32 Vertical bottom blasting renewal area versus ship's dimensions

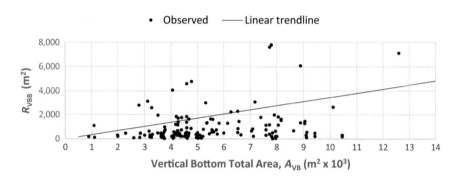

Fig. 7.33 Vertical bottom blasting renewal area versus vertical bottom total area

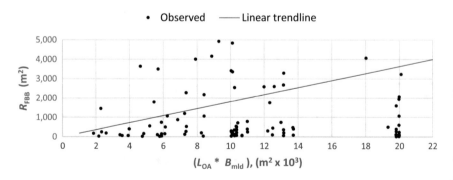

Fig. 7.34 Flat bottom blasting renewal area versus ship's dimensions

7.5.4 Hull Painting

Initial examination on pairs of interest variables related to the hull painting renewal area is shown in Figs. 7.38, 7.39, 7.40, 7.41, 7.42, 7.43, 7.44, 7.45, 7.46, 7.47,

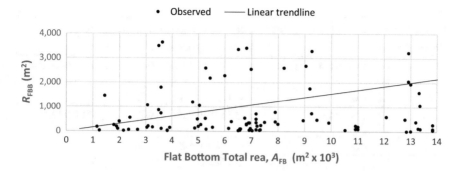

Fig. 7.35 Flat bottom blasting renewal area versus flat bottom total area

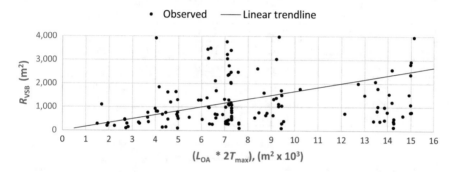

Fig. 7.36 Vertical side blasting renewal area versus ship's dimensions

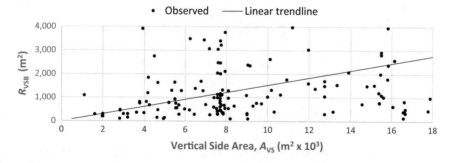

Fig. 7.37 Vertical side blasting renewal area versus vertical side total area

7.48 and 7.49. It displays the behaviour of the hull painting renewal area on the ship's dimensions and individual hull location area, respectively. All these figures reveal that with the increase of ships' dimensions, the painting area is also increased linearly. All these figures support the assumption made earlier that the size (length, breadth, and depth) of a ship positively impacts hull painting renewal area and is

Table 7.1 Summary of linear trendline equations and corresponding correlation coefficients of relationships between hull blasting renewal area, ship's dimensions, and actual area

Figure No	X	Y	Trendline equations	r^2	Locations
7.26	$[L_{OA} * (B_{mld} + 2 D_{mld})]/10^3$	R_{hb}	$Y = 128.110 * X$	0.874	Hull total area
7.27	$A_{HT}/10^3$	R_{hb}	$Y = 133.350 * X$	0.936	
7.28	$[L_{OA} * 2 (D_{mld} - T_{max})]/10^3$	R_{TSB}	$Y = 122.910 * X$	0.881	Topside total area
7.29	$A_{TS}/10^3$	R_{TSB}	$Y = 122.230 * X$	0.884	
7.30	$[L_{OA} * 2 (T_{max} - T_{min})]/10^3$	R_{BTB}	$Y = 178.460 * X$	0.940	Boottop total area
7.31	$A_{BT}/10^3$	R_{BTB}	$Y = 178.460 * X$	0.940	
7.32	$(L_{OA} * 2 T_{min})/10^3$	R_{VBB}	$Y = 366.100 * X$	0.827	Vertical bottom total area
7.33	$A_{VB}/10^3$	R_{VBB}	$Y = 344.810 * X$	0.785	
7.34	$(L_{OA} * B_{mld})/10^3$	R_{FBB}	$Y = 181.550 * X$	0.945	Flat bottom total area
7.35	$A_{FB}/10^3$	R_{FBB}	$Y = 154.240 * X$	0.911	
7.36	$(L_{OA} * 2 T_{max})/10^3$	R_{VSB}	$Y = 167.010 * X$	0.893	Vertical side total area
7.37	$A_{VS}/10^3$	R_{VSB}	$Y = 153.110 * X$	0.837	

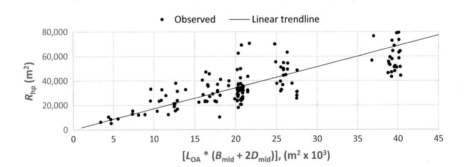

Fig. 7.38 Total painting renewal area versus ship's dimensions

linearly associated. A summary of linear trendline equations and the corresponding correlation coefficients of relationships between hull painting renewal area and ships' dimensions and actual areas is highlighted in Table 7.2.

Tables 7.1 and 7.2 suggest that the painting renewal areas have a stronger dependency on the ship's dimensions than blasting, which is very much expected. The probable reason is that the quantity of painting renewal area is truly governed by the hull area irrespective of the blasting renewal area. One may recall the painting process that blasted area is covered with one or two or more touch-up coats followed by one or two or more full coats to the entire area. Thus, the total painting area

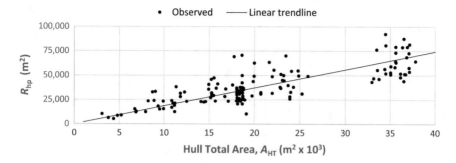

Fig. 7.39 Total painting renewal area versus hull total area

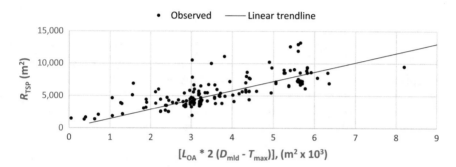

Fig. 7.40 Topside painting renewal area versus ship's dimensions

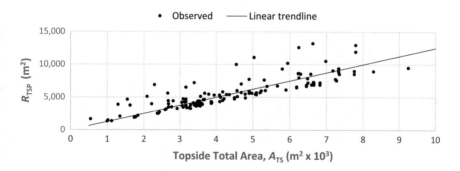

Fig. 7.41 Topside painting renewal area versus topside total area

depends on the hull total area. Total blasting area also depends on the hull total area but with a lower magnitude, reflected in lower constant slope value (m) than the painting renewal area. As such, the bigger ships (dimensionally) will require greater blasting and painting renewal works.

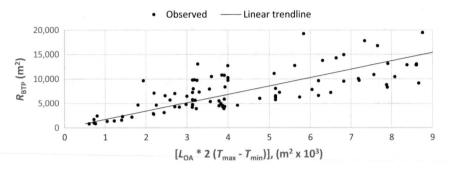

Fig. 7.42 Boottop painting renewal area versus ship's dimensions

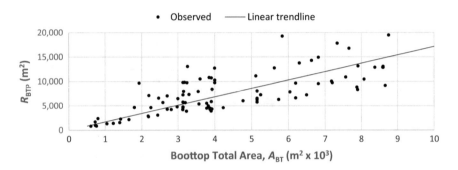

Fig. 7.43 Boottop painting renewal area versus boottop total area

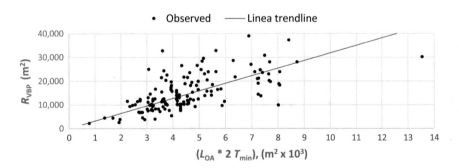

Fig. 7.44 Vertical bottom painting renewal area versus ship's dimensions

7.5.5   Hull Coating

Initial analysis of pairs of interest variables related to the hull coating renewal area is presented in Figs. 7.50 and 7.51. It depicts the behaviour of hull coating renewal area on ship's dimensions and the hull total area, respectively. Both figures reveal that with

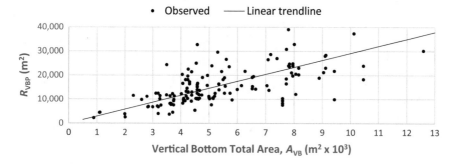

Fig. 7.45 Vertical bottom painting renewal area versus vertical bottom total area

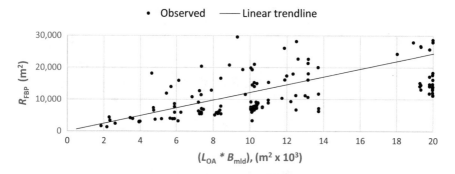

Fig. 7.46 Flat bottom painting renewal area versus ship's dimensions

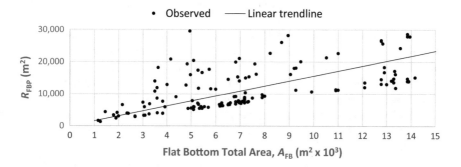

Fig. 7.47 Flat bottom painting renewal area versus flat bottom total area

the increase of ships' dimensions, the coating renewal area also increases linearly and supports the assumption made earlier that the size (length, breadth and depth) of a ship has a positive impact on hull coating renewal area and is linearly associated. A summary of linear trendline equations and the corresponding correlation coefficients

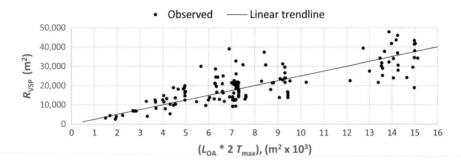

Fig. 7.48 Vertical side painting renewal area versus ship's dimensions

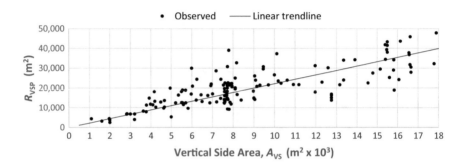

Fig. 7.49 Vertical side painting renewal area versus vertical side total area

of relationships between hull coating renewal area and ships' dimensions and actual areas is highlighted in Table 7.3.

Table 7.3 suggests that the coating renewal areas have a stronger dependency on the ship's dimensions than painting in terms of constant slope value, which is very much expected. The probable reason is that the quantity of coating renewal area has the combined effect of blasting and painting individually and is dominated by the painting area (80% and above). One may recall the painting process that blasted area is covered with one or two touch-up coats followed by one or two full coats to the entire area. Thus, the total coating area is higher than the painting area and depends on the hull total area. Total blasting area, painting area and coating area depend on ship's dimension and hull total area but with different responses and magnitude reflected different slope values (m). The blasting equation has the lowest, and the coating equation has the highest value for "m" (Table 7.4). As such, the bigger ships (dimensionally) will require greater blasting and painting renewal works.

Table 7.4 will be helpful to estimate the hull and location blasting and painting area at the preliminary stage using ship's dimensions and corresponding actual area if available.

Table 7.2 Summary of linear trendline equations and corresponding correlation coefficients of relationships between hull painting renewal area, ship's dimensions, and actual area

Figure No	X	Y	Trendline equations	r^2	Locations
7.38	$[L_{OA} * (B_{mld} + 2 *D_{mld})] /10^3$	R_{hp}	$Y = 1715.200 * X$	0.964	Hull total area
7.39	$A_{HT}/10^3$	R_{hp}	$Y = 1859.000 * X$	0.983	
7.40	$[L_{OA} * 2 (D_{mld} - T_{max})]/10^3$	R_{TSP}	$Y = 1444.400 * X$	0.963	Topside total area
7.41	$A^t_{TS}/10^3$	R_{TSP}	$Y = 1248.000 * X$	0.985	
7.42	$[L_{OA} * 2 (T_{max} - T_{min})]/10^3$	R_{BTP}	$Y = 1712.900 * X$	0.967	Boottop total area
7.43	$A_{BT}/10^3$	R_{BTP}	$Y = 1712.900 * X$	0.967	
7.44	$(L_{OA} * 2 T_{min})/10^3$	R_{VBP}	$Y = 3182.500 * X$	0.949	Vertical bottom total area
7.45	$A_{VB}/10^3$	R_{VBP}	$Y = 2913.000 * X$	0.941	
7.46	$(L_{OA} * B_{mld})/10^3$	R_{FBP}	$Y = 1219.400 * X$	0.880	Flat bottom total area
7.47	$A_{FB}/10^3$	R_{FBP}	$Y = 1556.600 * X$	0.876	
7.48	$(L_{OA} * 2 T_{max})/10^3$	R_{VSP}	$Y = 2504.300 * X$	0.975	Vertical side total area
7.49	$A_{VS}/10^3$	R_{VSP}	$Y = 2215.300 * X$	0.977	

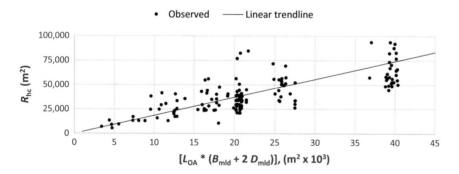

Fig. 7.50 Total coating renewal area versus ship's dimensions

7.6 Regression

In the previous sections and sub-sections, it has been highlighted that theoretically, age, deadweight and type are directly and positively associated with the corresponding hull coating (blasting and painting) renewal area of a ship. In other words, hull coating renewal area (dependent variable) is a function of age, deadweight, and type (independent variables). Mathematically, the relationships mentioned above may be expressed in equation form Eqs. (7.1)–(7.12).

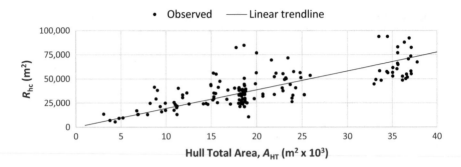

Fig. 7.51 Total coating renewal area versus hull total area

Table 7.3 Summary of linear trendline equations and corresponding correlation coefficients of relationships between hull coating renewal area, ship's dimensions, and actual area

Figure No	X	Y	Trendline equations	r^2	Locations
7.50	$[L_{OA} * (B_{mld} + 2 D_{mld})]/10^3$	R_{hc}	$Y = 1854.500 * X$	0.953	Hull total area
7.51	$A_{HT}/10^3$	R_{hc}	$Y = 1940.900 * X$	0.972	

$$R_{hb} = a + b * S_A \tag{7.1}$$

$$R_{hb} = a + b * S_D \tag{7.2}$$

$$R_{hb} = a + b * S_T \tag{7.3}$$

$$R_{hb} = a + b * (S_A * S_D) \tag{7.4}$$

$$R_{hp} = a + b * S_A \tag{7.5}$$

$$R_{hp} = a + b * S_D \tag{7.6}$$

$$R_{hp} = a + b * S_T \tag{7.7}$$

$$R_{hp} = a + b * (S_A * S_D) \tag{7.8}$$

$$R_{hc} = a + b * S_A \tag{7.9}$$

$$R_{hc} = a + b * S_D \tag{7.10}$$

Table 7.4 Summary of m for different linear trendline relationships between hull blasting, painting, and coating renewal area, ship's dimensions, and actual area

X	Y	Figure No	Trendline equations	Locations
$[L_{OA} * (B_{mld} + 2 D_{mld})]/10^3$	R_{hb}	26	$Y = 128.110 * X$	Hull total area
$A_{HT}/10^3$	R_{hb}	27	$Y = 133.350 * X$	
$[L_{OA} * 2 (D_{mld} - T_{max})]/10^3$	R_{TSB}	28	$Y = 122.910 * X$	Topside total area
$A_{TS}/10^3$	R_{TSB}	29	$Y = 122.230 * X$	
$[L_{OA} * 2 (T_{max} - T_{min})]/10^3$	R_{BTB}	30	$Y = 178.460 * X$	Boottop total area
$A_{BT}/10^3$	R_{BTB}	31	$Y = 178.460 * X$	
$(L_{OA} * 2 T_{min})/10^3$	R_{VBB}	32	$Y = 366.100 * X$	Vertical bottom total area
$A_{VB}/10^3$	R_{VBB}	33	$Y = 344.810 * X$	
$(L_{OA} * B_{mld})/10^3$	R_{FBB}	34	$Y = 181.550 * X$	Flat bottom total area
$A_{FB}/10^3$	R_{FBB}	35	$Y = 154.240 * X$	
$(L_{OA} * 2 T_{max})/10^3$	R_{VSB}	36	$Y = 167.010 * X$	Vertical side total area
$A_{VS}/10^3$	R_{VSB}	37	$Y = 153.110 * X$	
$[L_{OA} * (B_{mld} + 2 D_{mld})]/10^3$	R_{hp}	38	$Y = 1715.200 * X$	Hull total area
$A_{HT}/10^3$	R_{hp}	39	$Y = 1859.000 * X$	
$[L_{OA} * 2 (D_{mld} - T_{max})]/10^3$	R_{TSP}	40	$Y = 1444.400 * X$	Topside total area
$A_{TS}/10^3$	R_{TSP}	41	$Y = 1248.000 * X$	
$[L_{OA} * 2 (T_{max} - T_{min})]/10^3$	R_{BTP}	42	$Y = 1712.900 * X$	Boottop total area
$A_{BT}/10^3$	R_{BTP}	43	$Y = 1712.900 * X$	
$(L_{OA} * 2 T_{min})/10^3$	R_{VBP}	44	$Y = 3182.500 * X$	Vertical bottom total area
$A_{VB}/10^3$	R_{VBP}	45	$Y = 2913.000 * X$	
$(L_{OA} * B_{mld})/10^3$	R_{FBP}	46	$Y = 1219.400 * X$	Flat bottom total area
$A_{FB}/10^3$	R_{FBP}	47	$Y = 1556.600 * X$	
$(L_{OA} * 2 T_{max})/10^3$	R_{VSP}	48	$Y = 2504.300 * X$	Vertical side total area
$A_{VS}/10^3$	R_{VSP}	49	$Y = 2215.300 * X$	
$[L_{OA} * (B_{mld} + 2 D_{mld})]/10^3$	R_{hc}	50	$Y = 1854.500 * X$	Hull total area
$A_{HT}/10^3$	R_{hc}	51	$Y = 1940.900 * X$	

$$R_{hc} = a + b * S_T \tag{7.11}$$

$$R_{hc} = a + b * (S_A * S_D) \tag{7.12}$$

Since all the independent variables are found to be linearly associated with the dependent variable, so it is expected a multiple linear regression model will be an excellent fit for the system. A multiple linear regression model is considered to establish the relationship between hull coating renewal area, age, deadweight, and type. The following functions (Eqs. (7.13), (7.14) and (7.15)) are chosen because the hull coating renewal area is a function of each independent variable as per primarily mentioned assumptions.

$$R_{hb} = f(S_A, S_D, S_T) \tag{7.13}$$

$$R_{hp} = f(S_A, S_D, S_T) \tag{7.14}$$

$$R_{hc} = f(S_A, S_D, S_T) \tag{7.15}$$

Appropriate numerical values for S_T are calculated and assigned for types of ships for regression analysis. Table 7.5 displays the numerical values assigned to the types of ships for the functional equations (Eqs. (7.13), (7.14) and (7.15)). It is used in regression analysis to form Eqs. (7.16), (7.17) and (7.18).

The final regression equations are formed following the regression analysis method mentioned in Chapter 3 and using the observed data for R_{hb}, R_{hp}, R_{hc}, S_A, S_D, and S_T. They are as follows.

$$R_{hb} = -1683.630 + 189.024 * S_A + 0.007 * S_D + 324.180 * S_T \tag{7.16}$$

$$R_{hp} = 12602.450 + 395.000 * S_A + 0.150 * S_D + 433.060 * S_T \tag{7.17}$$

$$R_{hc} = 9948.310 + 579.300 * S_A + 0.150 * S_D + 909.830 * S_T \tag{7.18}$$

Table 7.5 Numerical values for types of ships for regression analysis

Types of ships	Numerical values for S_T for		
	Equation (7.13)	Equation (7.14)	Equation (7.15)
General cargo carrier	1.0000	1.0000	1.0000
Chemical tanker	2.9290	4.1264	4.0339
L.P.G. carrier	3.8670	4.7906	4.7193
Crude oil tanker	6.1071	7.4335	7.3311
Bulk carrier	8.2795	6.2558	6.4121
Container carrier	9.4416	6.1895	6.4407

Table 7.6 Summary of values of statistical parameters of final regression equations

Statistical parameters	Values of statistical parameters for		
	Equation (7.16)	Equation (7.17)	Equation (7.18)
Sample size (n)	143	143	143
No. of independent variable (k)	3	3	3
Significance level (α)	0.05	0.05	0.05
Standard deviation (s)	3229	11,906	14,040
Coefficient of determination (R^2)	0.143	0.591	0.536
$f - f_\alpha$	5.08	64.40	50.85

The vital statistical parameters regarding final regression Eqs. (7.16), (7.17) and (7.18) are given in Table 7.6. Improvement in coefficient of determination (R^2) is observed. The highest value is observed for Eq. (7.17). The condition $f > f_{0.05}$ (calculated F statistic and tabulated F statistic, respectively) suggested rejecting the null hypothesis. So, it may be concluded that there is a significant amount of variation in their response (the dependent variable) due to the differences in independent variables in the postulated models.

7.7 Validation

The validation technique is applied to Eqs. (7.16), (7.17) and (7.18). The summary of the validation results is presented in Table 7.7. The Table shows the outline of the variation of model values from the actual values. It is important to note that Eq. (7.16) yields hull blasting area, Eq. (7.17) yields hull painting area and Eq. (7.18) yields hull coating (blasting and painting jointly) area against age, deadweight, and type, respectively. The improvement compared to Eq. (7.16) is significant in terms of all parameters.

Table 7.7 Summary of validation results of final for regression equations

Items	Equation (7.16)	Equation (7.17)	Equation (7.18)
Positive error (%)	892	329	332
Negative error (%)	−91	−58	−61
Range of error (%)	983	387	393
Mean error (%)	105	13	14
Variance	33,492	2123	2205
Standard deviation	183	46	47

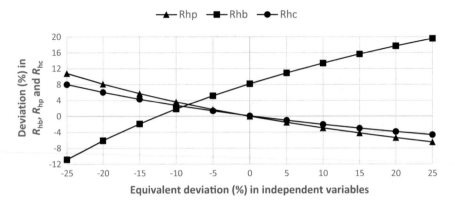

Fig. 7.52 Deviation in hull blasting renewal area, hull painting renewal area and hull coating renewal area versus equivalent deviation in independent variables

Reasons for deviation of model values from actual values explained in previous sections are equally applicable to the hull blasting, painting, and coating renewal area, but nature and magnitudes are different. Figure 7.52 demonstrates the relationship of deviation (%) in hull blasting renewal area, hull painting renewal area and hull coating renewal area against equivalent variation (%) in independent variables. The figure suggests that the hull painting and hull coating deviation follow the identical pattern, but hull blasting follows the opposite way. A similar phenomenon is observed in the validation result too. Equations (7.17) and (7.18) yields almost the same validation values. For Eqs. (7.17) and (7.18), equivalent negative deviation (%) in mean value will yield positive variation in model value, i.e., higher hull painting and coating area. But, for Eq. (7.16), it is the opposite. This behaviour entirely depends on the nature of the respective mathematical relationship (regression equations).

Validation is also applied to relationships of hull blasting (absolute value and percentage of hull total area) against age, deadweight and (age * deadweight) (Figs. 7.2, 7.4, 7.8, and Figs. 7.3, 7.5 and 7.9) respectively, hull painting (absolute value and percentage of hull total area) against age, deadweight and (age * deadweight) (Figs. 7.10, 7.12 and 7.16 and Figs. 7.11, 7.13 and 7.17) respectively and hull coating (absolute value and percentage of hull total area) against age, deadweight and (age * deadweight) (Figs. 7.18, 7.20 and 7.24 and Figs. 7.19, 7.21 and 7.25) respectively. Results are presented in Tables 7.8, 7.9, 7.10, 7.11, 7.12 and 7.13 respectively. It displays that validation result under painting and coating is better than that of under blasting in terms of the range of error, mean error, and standard deviation of error. It also indicates that the impact of independent variables is more significant for painting and coating.

The same technique is applied to relationships of hull blasting against ship's dimensions and hull total area (Figs. 7.26 and 7.27 respectively), hull painting against ship's dimensions and hull total area (Figs. 7.38 and 7.39, respectively) and hull

Table 7.8 Summary of validation results for blasting (m^2)

Items	Figure 7.2	Figure 7.4	Figure 7.8
Positive error (%)	1687	644	851
Negative error (%)	−85	−85	−94
Range of error (%)	1773	729	944
Mean error (%)	124	100	33
Variance	56,711	28,209	20,869
Standard deviation	238	168	144

Table 7.9 Summary of validation results for blasting (%)

Items	Figure 7.3	Figure 7.5	Figure 7.9
Positive error (%)	1207	918	900
Negative error (%)	−90	−83	−86
Range of error (%)	1298	1001	986
Mean error (%)	95	102	100
Variance	40,453	36,753	41,040
Standard deviation	201	192	203

Table 7.10 Summary of validation results for painting (m^2)

Items	Figure 7.11	Figure 7.13	Figure 7.17
Positive error (%)	924	321	480
Negative error (%)	−59	−60	−56
Range of error (%)	983	381	536
Mean error (%)	47	14	21
Variance	13,671	2434	5030
Standard deviation	117	49	71

Table 7.11 Summary of validation results for painting (%)

Items	Figure 7.10	Figure 7.12	Figure 7.15
Positive error (%)	90	90	87
Negative error (%)	−47	−46	−50
Range of error (%)	137	136	137
Mean error (%)	9	10	10
Variance	803	819	814
Standard deviation	28	29	29

coating against ship's dimensions and hull total area (Figs. 7.50 and 7.51 respectively). Results are presented in Tables 7.14, 7.15 and 7.16, respectively. It displays that validation result under painting and coating is better than that of under blasting in terms of the range of error, mean error, and standard deviation. It also indicates that the impact of independent variables is more significant for painting and coating.

Table 7.12 Summary of validation results for coating (m²)

Items	Figure 7.16	Figure 7.18	Figure 7.21
Positive error (%)	1079	357	516
Negative error (%)	−63	−63	−54
Range of error (%)	1142	420	569
Mean error (%)	47	18	28
Variance	15,983	2922	5604
Standard deviation	126	54	75

Table 7.13 Summary of validation results for coating (%)

Items	Figure 7.17	Figure 7.19	Figure 7.22
Positive error (%)	101	96	99
Negative error (%)	−53	−51	−55
Range of error (%)	155	147	154
Mean error (%)	5	11	11
Variance	966	975	971
Standard deviation	31	31	31

Table 7.14 Summary of validation results for blasting against dimensions

Items	Figure 7.26	Figure 7.27
Positive error (%)	882	848
Negative error (%)	−82	−82
Range of error (%)	964	931
Mean error (%)	116	103
Variance	37,824	34,322
Standard deviation	194	185

Table 7.15 Summary of validation results for painting against dimensions

Items	Figure 7.38	Figure 7.39
Positive error (%)	201	237
Negative error (%)	−51	−53
Range of error (%)	252	290
Mean error (%)	9	4
Variance	1092	1080
Standard deviation	33	33

Finally, the proposed mathematical models may be used to estimate the blasting, painting, and coating renewal area against various combinations of values for independent variables. Using the above model as a guide, shipyards may estimate the expected scope of hull blasting, painting area and the corresponding drydocking time, including logistics. While using the model to estimate the hull blasting and

Table 7.16 Summary of validation results for coating against dimensions

Items	Figure 7.50	Figure 7.51
Positive error (%)	216	250
Negative error (%)	−53	−59
Range of error (%)	270	310
Mean error (%)	11	5
Variance	1273	1256
Standard deviation	36	35

painting renewal area, one may be aware of the error level in predicting that. The model will provide a reasonable estimate of the expected independent variables that are close to the mean value. But for low and high value, the assessment will be low and high accordingly (Fig. 7.52). Also, one may consider allowing some allowance on top of the model value to accommodate the various variation of blasting and painting renewal areas explained earlier.

7.8 General Conclusions

The hull blasting and painting renewal area (in absolute value and percentage) reveal some very fundamental information regarding the hull total area, blasting area, painting area, and the coating repairing works of various types of ships. It breakdowns the average hull total area, blasting area and painting area against the hull locations like topside, boottop, vertical bottom and flat bottom. The contribution of individual locations in the hull total area, hull blasting area, and hull painting area is presented in Table 7.17 and Fig. 7.53.

The average hull total area consists of 21%, 19%, 25% and 36% for topside, boottop, vertical bottom and flat bottom, respectively, irrespective of age, deadweight and type. The average hull blasting work is about 12% of hull total area (A_{HT}), comprising 2%, 3%, 4% and 3% for topside, boottop, vertical bottom and flat bottom, respectively, irrespective of age, deadweight and type. The average hull painting work is 179% of the hull total area (A_{HT}), comprising 26%, 28%, 71% and 54% for

Table 7.17 Average area of location, blasting and painting against locations

Locations ↓	Average total area (%)		
	Hull	Blasting	Painting
Topside	20.61	1.90	26.20
Boottop	19.02	3.38	27.60
Vertical bottom	24.55	3.64	71.46
Flat bottom	35.82	3.38	54.11
Total	100	12.29	179.37

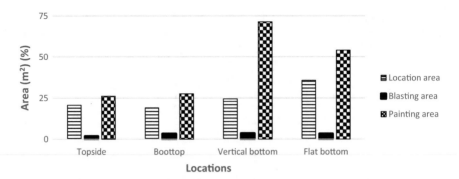

Fig. 7.53 Average location area, blasting area and painting area versus locations

topside, boottop, vertical bottom and flat bottom, respectively irrespective of age, deadweight and type of a ship. It is important to note that average blasting areas (%) against locations are always less than average painting areas (%) against locations. The reason is that blasting is always spot, meaning blasting area is less than hull total area but, painting is always multiple coats meaning that painting area is more than hull total area.

At this point, it is essential to note that whatever be the blasting area, there will be single or multiple touch-up coats to the blasted area followed by single or multiple full coats of the entire hull area. It explains why the total painting area (adding all painted area, touch up and full coats) is much more significant than a blasting area. It also explains why the total painting area is always higher than the hull total area (Figs. 7.27 and 7.41; Table 7.17). There are cases where the blasted area is very nominal compared to the hull total area, but the total painted area is more than two times the hull total area.

The analyses also found blasting and painting parameters for different ships and presented both tabular and graphical forms in Table 7.18 and Fig. 7.54, respectively. Container carriers claim the highest blasting and painting areas (%) and lowest for a crude oil tanker. These primary findings can be applied as a sort of proven guide for preliminary estimation purposes.

The analyses also reveal the relationships of hull surface area and its components with the ship's dimensions. They are described below in linear equation form and can preliminarily estimate hull area and its components irrespective of type.

Table 7.18 Average blasting and painting area against types of ships	Type of ship	Average area (%)	
		Blasting	Painting
	Crude oil tanker	12	179
	Chemical tanker	12	209
	LPG carrier	13	195
	Bulk carrier	22	198
	Container carrier	32	253

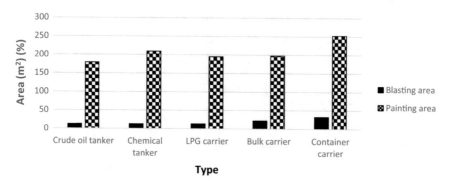

Fig. 7.54 Average hull blasting and painting area versus type

$$A_{\text{HT}} = 0.9022 * [L_{\text{OA}} * (B_{\text{mld}} + 2D_{\text{mld}})] \tag{7.19}$$

$$A_{TS} = 1.2014^* [L_{OA} * 2(D_{\text{mld}} - T_{\text{max}})] \tag{7.20}$$

$$A_{BT} = 1.00 * [L_{OA} * 2(T_{\text{max}} - T_{\text{min}})] \tag{7.21}$$

$$A_{\text{VB}} = 1.1564 * (L_{\text{OA}} * 2T_{\text{min}}) \tag{7.22}$$

$$A_{\text{FB}} = 0.6579 * (L_{\text{OA}} * B_{\text{mld}}) \tag{7.23}$$

$$A_{\text{VS}} = 1.1058 * (L_{\text{OA}} * 2T_{\text{max}}) \tag{7.24}$$

Figure 7.55 is developed using the relationships determined in Figs. 7.3, 7.11 and 7.18 between hull blasting, painting and coating renewal area against age, respectively. It displays the estimated hull blasting, painting, and coating renewal areas (%) against age irrespective of deadweight and type.

Figure 7.56 is developed using the relationships determined in Figs. 7.5, 7.13 and 7.21 between hull blasting, painting, and coating renewal area against deadweight, respectively. It displays the estimated hull blasting, painting, and coating renewal areas (%) against deadweight irrespective of age and type.

Figure 7.57 is developed using the relationships determined in Figs. 7.9, 7.17 and 7.25 between hull blasting, painting, and coating renewal area against (age * deadweight), respectively. In addition, it displays the estimated hull blasting, painting, and coating renewal areas (%) against (age * deadweight) irrespective of type.

Figures 7.58, 7.59 and 7.60 are developed by expanding Fig. 7.7 using the same sample data. It displays the estimated hull blasting renewal area (%) against age, deadweight, and age x deadweight, respectively, for different types.

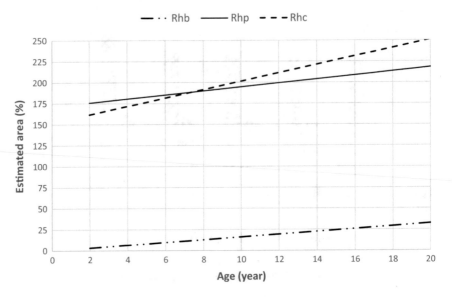

Fig. 7.55 Estimated hull blasting, painting, and coating renewal area versus age

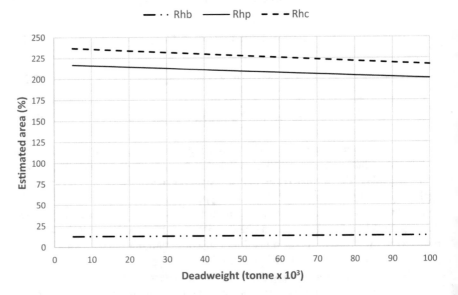

Fig. 7.56 Estimated hull blasting, painting, and coating renewal area versus deadweight

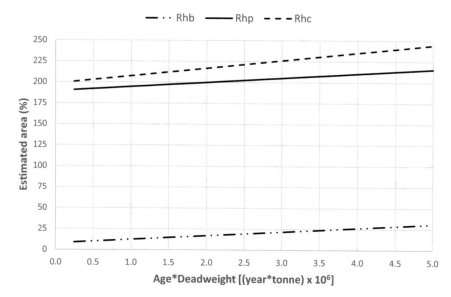

Fig. 7.57 Estimated hull blasting, painting, and coating renewal area versus (age * deadweight)

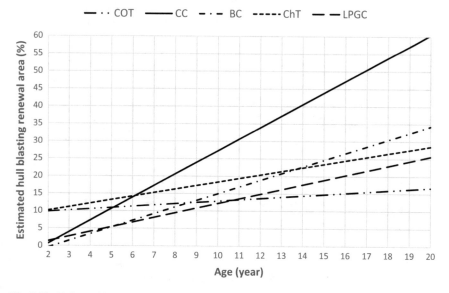

Fig. 7.58 Estimated hull blasting renewal area versus age for crude oil tankers, container carriers, bulk carriers, chemical tankers and liquified petroleum gas carriers

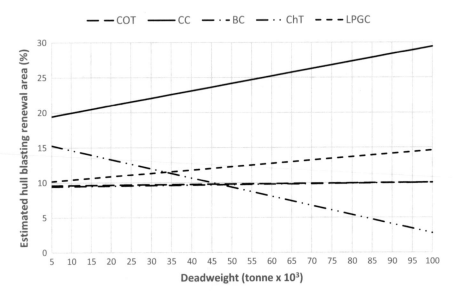

Fig. 7.59 Estimated hull blasting renewal area versus deadweight for crude oil tankers, container carriers, bulk carriers, chemical tankers and liquified petroleum gas carriers

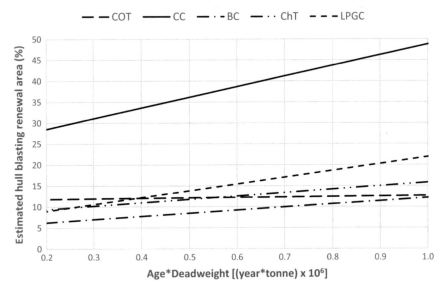

Fig. 7.60 Estimated hull blasting renewal area versus (age * deadweight) for crude oil tankers, container carriers, bulk carriers, chemical tankers and liquified petroleum gas carriers

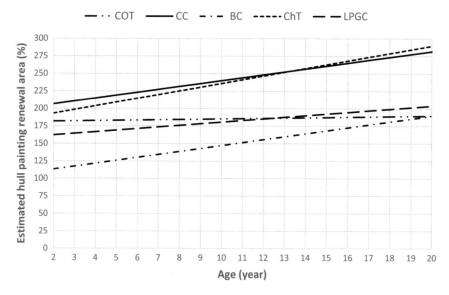

Fig. 7.61 Estimated hull painting renewal area versus age for crude oil tankers, container carriers, bulk carriers, chemical tankers and liquified petroleum gas carriers

Figures 7.61, 7.62 and 7.63 are developed by expanding the Fig. 7.15 using the same sample data. It displays the estimated hull painting renewal area (%) against age, deadweight and (age * deadweight), respectively, for different types.

Figures 7.64, 7.65 and 7.66 are developed by expanding the Fig. 7.23 using the same sample data. It displays the estimated hull coating renewal area (%) against age, deadweight, and age x deadweight, respectively, for different types.

Estimating hull blasting and painting renewal areas can be carried out using the respective figure and related variables. One may also follow the below options with various independent variables to estimate blasting and painting renewal areas.

Option—I
Use age and estimate hull blasting and painting renewal area (%) irrespective of deadweight and type with the help of Fig. 7.55.

Option—II
Use deadweight and estimate hull blasting and painting renewal area (%) irrespective of age and type with the help of Fig. 7.56.

Option—III
Use (age * deadweight) and estimate hull blasting and painting renewal area (%) irrespective of type with the help of Fig. 7.57.

Option—IV
Use age and type and estimate the hull blasting renewal area (%) for the corresponding type, irrespective of deadweight, with the help of Fig. 7.58 as appropriate.

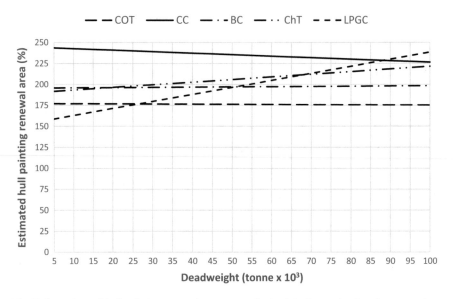

Fig. 7.62 Estimated hull painting renewal area versus deadweight for crude oil tankers, container carriers, bulk carriers, chemical tankers and liquified petroleum gas carriers

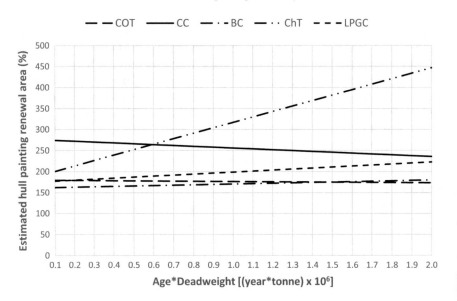

Fig. 7.63 Estimated hull painting renewal area versus (age * deadweight) for crude oil tankers, container carriers, bulk carriers, chemical tankers and liquified petroleum gas carriers

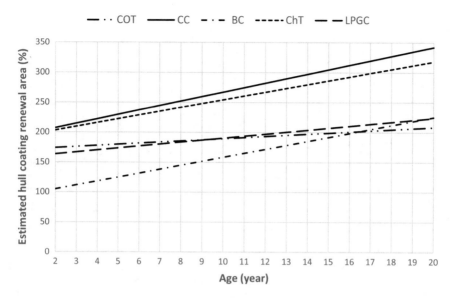

Fig. 7.64 Estimated hull coating renewal area versus age for crude oil tankers, container carriers, bulk carriers, chemical tankers and liquified petroleum gas carriers

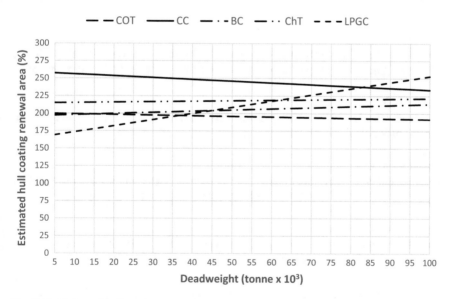

Fig. 7.65 Estimated hull coating renewal area versus deadweight for crude oil tankers, container carriers, bulk carriers, chemical tankers and liquified petroleum gas carriers

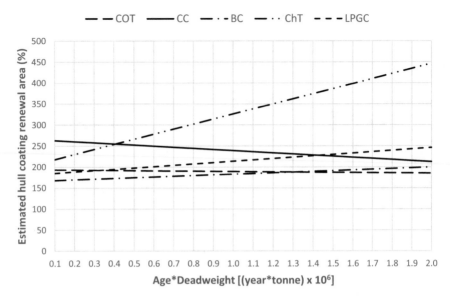

Fig. 7.66 Estimated hull coating renewal area versus (age * deadweight) for crude oil tankers, container carriers, bulk carriers, chemical tankers and liquified petroleum gas carriers

Option—V
Use deadweight and type and estimate hull blasting renewal area (%) for the corresponding type, irrespective of age, with the help of Fig. 7.59 as appropriate.

Option—VI
Use (age * deadweight) and the type and estimate hull blasting renewal area (%) for the corresponding type with the help of Fig. 7.60 as appropriate.

Option—VII
Use age and type and estimate hull painting renewal area (%) for the corresponding type, irrespective of deadweight, with the help of Fig. 7.61 as appropriate.

Option—VIII
Use deadweight and type and estimate hull painting renewal area (%) for the corresponding type, irrespective of age, with the help of Fig. 7.62 as appropriate.

Option—IX
Use (age * deadweight) and type and estimate hull painting renewal area (%) for the corresponding type with the help of Fig. 7.63 as appropriate.

Option X
Use age, deadweight and type and estimate hull blasting renewal area (m²) for the corresponding type, with the help of regression Eq. (7.16).

Option—XI

Use age, deadweight and type and estimate hull painting renewal area (m^2) for the corresponding type with the help of regression Eq. (7.17).

Figures 7.53 and 7.54 may be used to estimate blasting and painting renewal area (%) for types irrespective of age and deadweight.

Hull total area (A_{HT}) and Individual hull location areas (A_{TS}, A_{BT}, A_{VB}, A_{FB} and A_{VS}) also may be estimated using ship's dimensions with the help of Eqs. (7.19)–(7.24), respectively.

Individual hull location blasting, and painting area (m^2) may be estimated using the appropriate linear equation with the help of Table 7.4.

Finally, it is up to the individuals to use the findings according to their requirements.

References

1. Broderick, D.R., Wright, P.N.H., Kattan, M.R.: Exploring the link between structural complexity and coating performance. In: Proceeding, 11th International Marine Design Conference (IMDC), June 11–14, Glasgow, UK (2012)
2. Jotun, C.M.: Chapter 12 Surface Preparation and Cleaning, 12.42 (2001)
3. Dev, A.K., Saha, M.: Analysis of hull coating renewal in ship repairing. J. Ship Prod. Des. **33**(3), 197–211 (2017)
4. Garbatov, Y., Soares, C.G., Wang, G.: Non-linear time-dependent corrosion wastage of deck plate of ballast and cargo tanks of tankers. Am. Soc. Mech. Eng. (ASME) **129**(1), 48–55 (2006)
5. Hiromi, S., Matsushita, H., Song, Y., Nakai, T., Yuya, N.: Thickness reduction due to flow accelerated corrosion of shipboard piping. ClassNK Tech. Bull. 35–56 (2006)
6. Nakai, T., Matsushita, H., Yamamoto, N.: Assessment of corroded conditions of webs of hold frames with pitting corrosion. ClassNK Tech. Bull. 23–32 (2007)
7. O'Donnell, W.J.: Corrosion fatigue: recent developments and future needs. J. Marine Des. Oper. 27–37 (2006)
8. Paik, J.K., Thayamballi, A.K., Kim, S.K., Yang, S.H.: Ship hull ultimate strength reliability considering corrosion. J. Ship Res. **42**(2), 154–165 (1998)
9. Yamamoto, N., Ikegami, K.: A study on the degradation of coating and corrosion of ship's hull based on the probabilistic approach. J. Offshore Mech. Arct. Eng. **120**, 121–128 (1998)

Chapter 8
Structural Steel Renewal Weight

8.1 Introduction

Structural steel repair/replacement of a ship occurs under two severe conditions, such as (i) reduction of the thickness of structural members beyond the limit set by the classification societies and (ii) deformation of structural members beyond the limit set by the classification societies. In general, a reduction in the thickness of a structural steel member occurs due to many reasons. These are general corrosion phenomena commonly taking place on the un-coated steel surface, grooving corrosion often found in heat affected areas, pitting corrosion frequently appearing due to local coating breakdown, corrosion due to fluid flow by mechanical action and galvanic corrosion (when dissimilar materials come in contact with water, especially seawater). Other causes are the oxidation process (when exposed steel members contact oxygen in the air), natural wear and tear, mechanical contact, etc. Deformation may also occur due to physical ship movement (hogging, sagging, rolling, torsion, combined movement, etc.), contact damage (collision, grounding, hit by the underwater object), etc. Whatever the reason for structural steel renewal, the main reason is the structural members' deficiencies beyond the limit, either regarding scantling (thickness) or structural deformations (dimensional) or both.

Structural steel renewal weight has a significant impact on the ship repairing budget. It is also an accepted fact that the labour and structural steelworks related costs are the two main components of ship repairing cost [3]. Therefore, preliminary information about the scope of structural steel repairing works will help the shipowners to allocate an appropriate budget and prepare a cargo schedule to meet the commercial commitment. It is also true that the structural steel repairing work consumes maximum resources from the shipyard side. Therefore, preliminary information will help the shipyards prepare a realistic schedule by optimising resource allocation and utilisation. But there is no tool, in terms of the guideline, to calculate the expected structural steel renewal weight against prevalent variables like age, deadweight and type.

A. K. Dev et al., *Ship Repairing*, Springer Series on Naval Architecture,
Marine Engineering, Shipbuilding and Shipping 12,
https://doi.org/10.1007/978-981-16-9468-4_8

No academic paper, article or research work has been found devoted to structural steel renewal of a ship concerning their age, deadweight, and type. The probable reason seems to be the scarcity and confidentiality of such classified information and data. However, some works, not exactly but close to the issue, were done from different viewpoints. Dev and Saha [1] investigated structural steel renewal weight during the routine maintenance schedule. This article attempts to demonstrate the trends of the quantity of structural steel renewal in ship repairing concerning age, deadweight, CN, GT, and type of ships. This data analysis suggests that the structural steel renewal works, and their components are a function of age, deadweight, CN, GT, and type, but at different degrees of responses. It also reveals some fundamental basis for estimating normal structural steel renewal scope for various ages, deadweights, CNs, GTs, and types. All independent variables are mostly linearly associated with the dependent variable. Seref and Mirza [4] examined structural steel renewal works of bulk carriers. The paper demonstrates the locations and trends of repairs caused by factors relating to ships' operations, exposure to natural elements of their operating environment at sea, and quality of ship staff, all of which point to a predictive pattern where repairs are most likely to occur bulk carriers. Hiromi et al. [2] analysed the reduction in wall thickness of shipboard various system pipes, due to flow-accelerated corrosion (FAC), under different flow conditions and pipe geometry. They also proposed using Kastner's experimental formula to estimate the reduction in wall thickness of pipelines onboard. Yamamoto et al. [5] studied the corrosion phenomenon of a ship's hull consisting of three sequential processes such as (i) degradation of paint coatings, (ii) generation of pitting points and (iii) progress of pitting point. They described each process by introducing a probabilistic corrosion model. This probabilistic corrosion model can be developed by analysing existing data collected from plate thickness measurements. By comparing the results of estimations by the identified probabilistic models and actual measurement data, the practical usefulness of the proposed procedure is proved.

This Chapter will discuss how the age, deadweight, type separately and together influence the total structural steel renewal weight. In this regard, there will be some assumptions and subsequently, verify them through analysing the respective variable. In this analysis, total structural steel renewal weight is considered a dependent variable, and others thought independent variables. Finally, a mathematical model is presented to predict the expected weight of total structural steel to be renewed against a set of independent variables like age, deadweight, and type. It also added graphs to estimate structural steel renewal weight and components' renewal weight against various combinations of independent variables.

For simplification and easy to understand, the analysis of **total structural steel renewal weight** is divided into four major groups as follows:

(1)   Total hull plates renewal weight
(2)   Total transverse structural members' renewal weight
(3)   Total longitudinal structural members' renewal weight
(4)   Total miscellaneous steel renewal weight.

For further simplification, each group is divided into various subgroups (by members) as follows.

Hull plates (from now on also referred to as a plate group)

(1) Bottom plates
(2) Shell plates
(3) Deck plates
(4) Tank top plates.

Transverse structural members (from now on also referred to as a transverse group)

(1) Bottom transverses
(2) Deck transverses
(3) Transverse bulkheads including stiffeners
(4) Side frames.

Longitudinal structural members (from now on also referred to as a longitudinal group)

(1) Longitudinal bulkheads
(2) Longitudinal girders
(3) Bottom longitudinals
(4) Tank top longitudinals
(5) Shell longitudinals
(6) Deck longitudinals
(7) Longitudinal bulkhead longitudinals
(8) Side stringers.

Miscellaneous structural members (from now on referred to as a miscellaneous group)

(1) Bilge keels
(2) Brackets
(3) Carlings
(4) Lug pieces
(5) Collar plates
(6) Stiffeners.

A general assumption is that age, deadweight, type, and (age * deadweight) separately impact the structural steel renewal weight and its linearly related components. Therefore, in the following sections, each of them will be discussed.

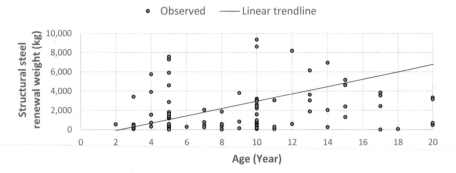

Fig. 8.1 Total structural steel renewal weight versus age

8.2 Total Structural Steel Renewal Weight Versus Age (R_S vs S_A)

As per general assumption, age positively impacts structural steel replacement weight irrespective of deadweight and type and linearly associated. It means that older ships will demand more structural steel renewal compared to newer ones.

Initial investigation of structural steel renewal weight versus age is demonstrated in Fig. 8.1, which shows the total structural steel renewal weight against age irrespective of deadweight and type. It offers a positive relationship. It is very likely because as ships become older, the scantlings of structural members are wasted and reduced from original dimensions due to the different unavoidable natural phenomena, which ultimately leads to the member's replacement when reduction exceeds the limit. The linear equation, $R_S = -852.810 + 380.590 * S_A$, provides the best goodness of fit to sample data with a correlation coefficient of 0.791.

Therefore, the assumption made is valid. However, more clearly, older ships are expected to have more structural steel renewal works than newer ones.

8.3 Total Structural Steel Renewal Weight Versus Deadweight (R_S vs S_D)

General assumption suggests that deadweight directly influences the structural steel renewal weight with a linear association. Hence, the structural steel renewal weight is a function of deadweight irrespective of age and type and linearly associated.

Examination of structural steel renewal weight versus deadweight is presented in Fig. 8.2, which shows structural steel renewal weight against deadweight irrespective of age and type. It offers a positive relationship. It is very much evident because a higher deadweight means dimensionally larger size, meaning bigger structural steel. Thus, logically the bigger ships will have more extensive structural steel replacements

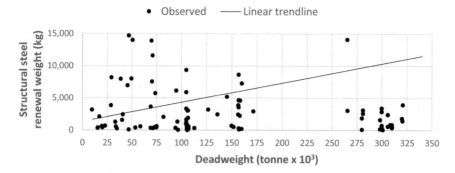

Fig. 8.2 Total structural steel renewal weight versus deadweight

works. The linear equation, $R_S = 1331.500 + 30.027 * (S_D/10^3)$, yields the best goodness of fit to sample data with a correlation coefficient of 0.688.

Therefore, the assumption is valid, so bigger ships will see more structural steel renewal works than smaller ones.

8.4 Total Structural Steel Renewal Weight Versus Type (R_S vs S_T)

It is assumed that type influences the structural steel renewal weight and varies linearly irrespective of age and deadweight.

Analysis of structural steel renewal weight against type is highlighted in Fig. 8.3, which shows structural steel renewal weight behaviour. The figure shows a strong relationship. It is expected due to the inherent differences of types and their operational and performance requirements.

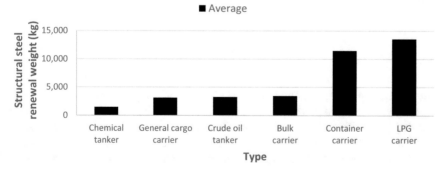

Fig. 8.3 Average total structural steel renewal weight versus type

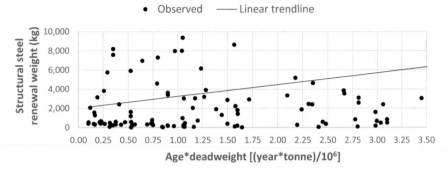

Fig. 8.4 Total structural steel renewal weight versus (age * deadweight)

Therefore, the assumption is valid. Different ships will have various structural steel renewals even though they are of the same age and size.

8.5 Total Structural Steel Renewal Weight Versus (age * deadweight) [R_S vs ($S_A * S_D$)]

Structural steel renewal weight is assumed to be a function of (age * deadweight) and linearly involved. Studies of structural steel renewal weight and (age * deadweight) are displayed in Fig. 8.4 showing structural steel renewal weight against (age * deadweight) irrespective of type. The figure shows a positive relationship. This is logical because the higher (age * deadweight) values demand a bigger size or older ship and, eventually, higher structural steel renewal weight. The linear equation, $R_S = 2015.100 + 1230.700 * [(S_A * S_D)/10^6]$, delivers the best goodness of fit to sample data with a correlation coefficient of 0.194.

Therefore, the assumption made is valid. So, bigger and older ships are expected to have more structural steel renewal works than smaller and newer ships.

8.6 Plate Group

The plate group covers deck plates, side shell plates, tank top plates and bottom plates (also often called skin plates). They are the prominent structural members of a ship and, together with stiffening members (transverse and longitudinal), significantly contribute to the longitudinal strength like a considerable box-girder. Moreover, each plate acts against unavoidable natural action when a ship moves in addition to structural strength.

Deck and bottom plates, alternately, experience tensile and compressive loads together with torsional effects due to the ship's motions, like hogging, sagging, rolling, pitching, heaving, etc. Side shell plates withstand the hydrostatic pressure and wave forces when a ship is at rest or in motion. As a result, hydrostatic pressure and wave load vary with the ship's motions and draft. Tank top plates (cargo tank/cargo hold bottom plates) experience the same stress phenomena as deck and bottom plates. In addition, it bears the static load due to cargo loaded. Bottom shell plates withstand hydrostatic pressure and vertical wave forces. As a result, hydrostatic pressure and wave load vary with the ship's motions and draft.

The last few sections establish that structural steel renewal weight independently is a function of age, deadweight, type and (age * deadweight), and they are linearly associated. Therefore, it is also logical that components of structural steel renewal weight will behave the same way. More clearly, plate, transverse, longitudinal, and miscellaneous groups are independently a function of the variables mentioned above with linear association. Therefore, all structural group renewal weight will be discussed following a similar assumption in the following sections.

8.6.1 Plate Group Renewal Weight Versus Age (R_{PG} vs S_A)

The explanation of plate group renewal weight versus age is depicted in Fig. 8.5, which shows the behaviour of plate group renewal weight against age irrespective of deadweight and type and shows a fair relationship. The linear equation, $R_{PG} = 104.600 + 110.150 * S_A$, forecasts the best goodness of fit relationship with a correlation coefficient of 0.527.

Therefore, the assumption is valid as such. Older ships are expected to see more plate group renewal weight than newer ones.

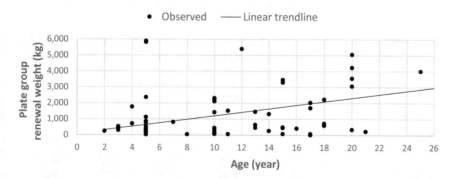

Fig. 8.5 Plate group renewal weight versus age

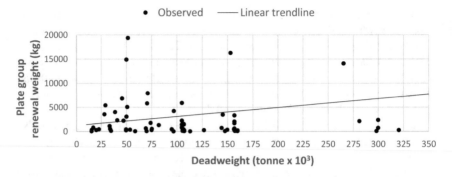

Fig. 8.6 Plate group renewal weight versus deadweight

8.6.2 Plate Group Renewal Weight Versus Deadweight (R_{PG} vs S_D)

Evaluation of plate group renewal weight against deadweight is produced in Fig. 8.6, which shows the plate group renewal weight against deadweight irrespective of age and type. It offers a positive relationship. The linear equation, $R_{PG} = 1243.100 + 18.589 * (S_D/10^3)$, predicts the best goodness of fit relationship with a correlation coefficient of 0.159.

Therefore, the assumption made is valid. More specifically, bigger ships are likely to demand more plate group renewal works than smaller ones.

8.6.3 Plate Group Renewal Weight Versus Type (R_{PG} vs S_T)

Initial investigation of plate group renewal weight versus type is exhibited in Fig. 8.7, showing a linear relationship. Therefore, the assumption is valid. Hence, different

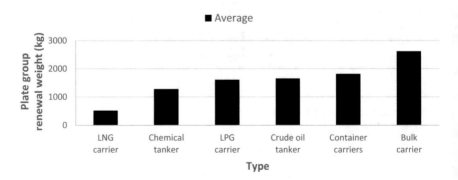

Fig. 8.7 Average plate group renewal weight versus type

types of ships will have other plate group renewal amounts even if they are of the same age and deadweight.

8.6.4 Plate Group Renewal Weight Versus (age * deadweight) [R_{PG} vs (S_A * S_D)]

Initial examination of structural steel renewal weight and (age * deadweight) is demonstrated in Fig. 8.8, which shows the plate group renewal weight against (age * deadweight) irrespective of type. It shows a positive relationship. The linear equation, $R_{PG} = 1141.300 + 4214.300 * [(S_A * S_D)/10^6]$, estimates the best goodness of fit to sample data with a correlation coefficient of 0.969.

Therefore, the assumption made is valid. As such, bigger and older ships are likely to have more plate group renewal weight than smaller and newer ships.

8.7 Transverse Group

The transverse group covers bottom transverses, deck transverses, transverse bulk-heads, including stiffeners and side frames attached to the side shell and longitudinal bulkhead. Together with skin plate, all these members act against various forces in various operational and environmental conditions.

The transverse section of a ship is always subjected to static pressure (hydrostatic) due to the surrounding water and subjected to internal loading due to the weight of the internal structure itself and its weight. The effects of these static forces are to cause transverse distortion of the section. It is worth stating that this type of distortion is independent of longitudinal bending (in a vertical plane and horizontal plane).

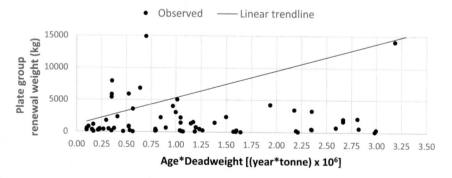

Fig. 8.8 Plate group renewal weight versus (age * deadweight)

8.7.1 Transverse Group Renewal Weight Versus Age (R_{TG} vs S_A)

Analysis of transverse group renewal weight against age is presented in Fig. 8.9, which shows the behaviour of transverse group renewal weight against age irrespective of deadweight and type and shows a fair and positive relationship. The linear equation, $R_{TG} = -574.990 + 167.290 * S_A$, provides the best goodness of fit to sample data with a correlation coefficient of 0.793.

Therefore, the assumption made is valid. However, more clearly, older ships are likely to see more transverse group renewal weight than newer ones.

8.7.2 Transverse Group Renewal Weight Versus Deadweight (R_{TG} vs S_D)

The initial study of transverse group renewal weight versus deadweight is highlighted in Fig. 8.10, which shows the behaviour of total transverse group renewal weight against deadweight irrespective of age and type. It offers a positive relationship. The linear equation, $R_{TG} = 316.570 + 2.591 * (S_D/10^3)$, yields the best goodness of fit to sample data with a correlation coefficient of 0.438.

However, the assumption is valid so that bigger ships will see more transverse group renewal weight than smaller ones.

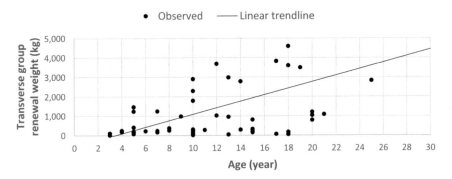

Fig. 8.9 Transverse group renewal weight versus age

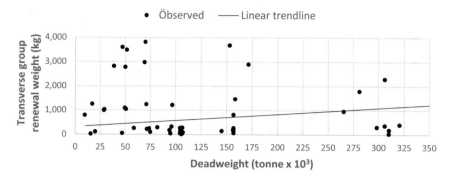

Fig. 8.10 Transverse group renewal weight versus deadweight

8.7.3 Transverse Group Renewal Weight Versus Type (R_{TG} vs S_T)

Initial exploration of transverse group renewal weight versus type is exhibited in Fig. 8.11. It shows a linear relationship. Therefore, the assumption made is valid. It means different ships will have an additional transverse group renewal even if they are of the same age and deadweight.

8.7.4 Transverse Group Renewal Weight Versus (age * deadweight) [R_{TG} vs (S_A * S_D)]

Initial review of transverse group renewal weight and (age * deadweight) is displayed in Fig. 8.12, which shows the transverse group renewal weight against (age * deadweight) irrespective of type. Figures show a positive relationship. The linear equation,

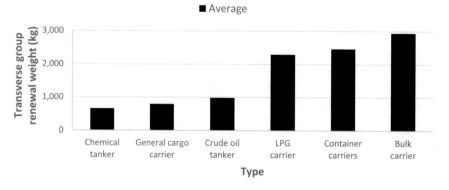

Fig. 8.11 Average transverse group renewal weight versus type

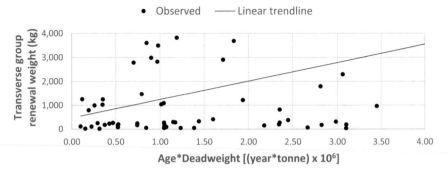

Fig. 8.12 Transverse group renewal weight versus (age * deadweight)

$R_{TG} = 475.530 + 768.990 * [(S_A * S_D)/10^6]$, delivers the best goodness of fit with a correlation coefficient of 0.636.

Therefore, the assumption made is valid. Hence, bigger and older ships are anticipated to have more transverse group renewal weight than smaller and newer ships.

8.8 Longitudinal Group

The longitudinal group covers longitudinal bulkheads, longitudinal girders, bottom longitudinals, tank top longitudinals, shell longitudinals, deck longitudinals, longitudinal bulkhead longitudinals and side stringers. Together with skin plate, all these members act against various forces in various operational and environmental conditions like a transverse group.

8.8.1 Longitudinal Group Renewal Weight Versus Age (R_{LG} vs S_A)

Initial evaluation of longitudinal group renewal weight versus age is depicted in Fig. 8.13, which shows the behaviour of longitudinal group renewal weight against age irrespective of deadweight and type. It offers an honest relationship. The linear equation, $R_{LG} = 542.040 + 238.390 * S_A$, forecasts the best goodness of fit relationship with a correlation coefficient of 0.813.

However, the assumption made is valid. More clearly, older ships are expected to see more longitudinal group renewal weight than newer ones.

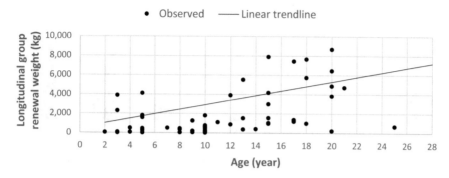

Fig. 8.13 Longitudinal group renewal weight versus age

8.8.2 *Longitudinal Group Renewal Weight Versus Deadweight (R_{LG} vs S_D)*

Initial investigation of longitudinal group renewal weight versus deadweight is introduced in Fig. 8.14, which shows the behaviour of total longitudinal group renewal weight against deadweight irrespective of age and type. It offers an honest and positive relationship. The linear equation, $R_{LG} = -163.070 + 19.495 * (S_D/10^3)$, predicts the best goodness of fit relationship with a correlation coefficient of 0.396.

Therefore, the assumption made is valid. Mainly, bigger ships are supposed to see more longitudinal group renewal weight than smaller ones.

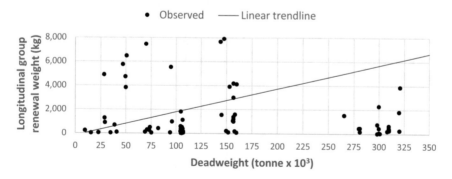

Fig. 8.14 Longitudinal group renewal weight versus deadweight

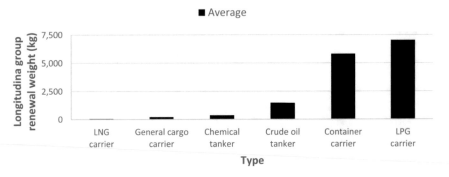

Fig. 8.15 Average longitudinal group renewal weight versus type

8.8.3 Longitudinal Group Renewal Weight Versus Type (R_{LG} vs S_T)

Initial examination of longitudinal group renewal weight against type is demonstrated in Fig. 8.15, showing a linear relationship. Therefore, the assumption made is valid, and obviously, different types of ships will have an additional amount of longitudinal group renewal even they are of the same age and deadweight.

8.8.4 Longitudinal Group Renewal Weight Versus (age * deadweight) [R_{LG} vs ($S_A * S_D$)]

Analysis of longitudinal group renewal weight and (age * deadweight) are presented in Fig. 8.16, which shows longitudinal group renewal weight behaviour against (age * deadweight) irrespective of type. It shows a positive relationship. The linear

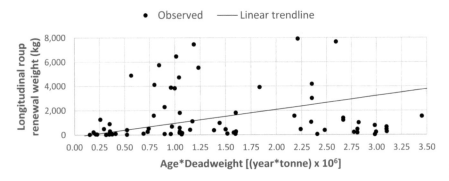

Fig. 8.16 Longitudinal group renewal weight versus (age x deadweight)

equation, $R_{LG} = -176.850 + 1131.200 * [(S_A * S_D)/10^6]$, estimates the best goodness of fit to system with a correlation coefficient of 0.598.

Therefore, the assumption made is valid. So, bigger and older ships are assumed to have more longitudinal group renewal weight than smaller and newer ships.

8.9 Miscellaneous Group

The miscellaneous group covers bilge keels, brackets, carlings, lug pieces, collar plates and stiffeners. All these members, except bilge keel—used as an anti-rolling device—contribute to the structural strength (transverse and longitudinal).

8.9.1 Miscellaneous Group Renewal Weight Versus Age (R_{MS} vs S_A)

The study of miscellaneous group renewal weight versus age is highlighted in Fig. 8.17, which shows the behaviour of total miscellaneous group renewal weight against age irrespective of deadweight and type. It offers a positive relationship. The linear equation, $R_{MS} = 3.122 + 66.706 * S_A$, provides the best goodness of fit relationship with a correlation coefficient of 0.694.

However, the assumption made is valid. More particularly, older ships are expected to see more miscellaneous group renewal weight than newer ones.

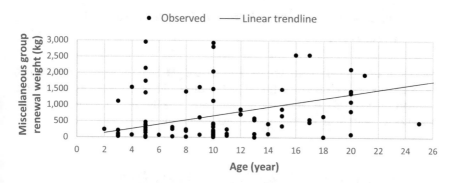

Fig. 8.17 Miscellaneous group renewal weight versus age

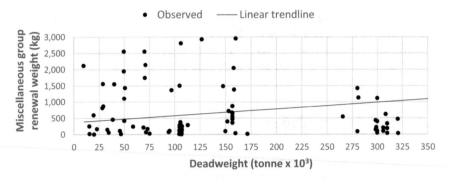

Fig. 8.18 Miscellaneous group renewal weight versus deadweight

8.9.2   Miscellaneous Group Renewal Weight Versus Deadweight (R_{MS} vs S_D)

Initial review of miscellaneous group renewal weight versus age is displayed in Fig. 8.18 showing the total miscellaneous group renewal weight against deadweight irrespective of age and type. It offers an honest relationship. The linear equation, $R_{MS} = 370.070 + 2.070 * (S_D/10^3)$, yields the best goodness of fit relationship with a correlation coefficient of 0.354.

However, the assumption made is valid. It means bigger ships are anticipated to see more miscellaneous group renewal weight than smaller ones.

8.9.3   Miscellaneous Group Renewal Weight Versus Type (R_{MS} vs S_T)

Evaluation of miscellaneous group renewal weight against type is depicted in Fig. 8.19. It shows a linear relationship. Therefore, the assumption made is valid. Different types of ships will have an additional amount of miscellaneous group renewal even if they are of the same age and deadweight.

8.9.4   Miscellaneous Group Renewal Weight Versus (age * deadweight) [R_{MS} vs (S_A * S_D)]

Investigation of miscellaneous group renewal weight and (age * deadweight) is introduced in Fig. 8.20, which shows the behaviour of the miscellaneous group renewal

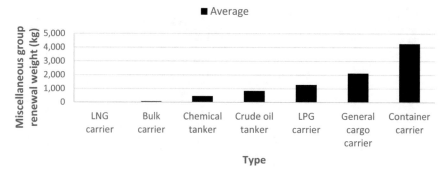

Fig. 8.19 Average miscellaneous group renewal weight versus type

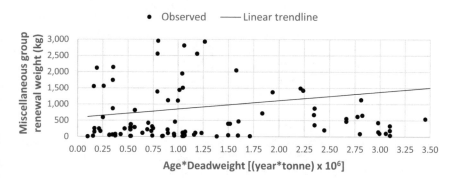

Fig. 8.20 Miscellaneous group renewal weight versus (age * deadweight)

weight against (age * deadweight) irrespective of type. It offers an honest relationship. The linear equation, $R_{MS} = 587.170 + 261.600 * [(S_A * S_D)/10^6]$, delivers the best goodness of fit to sample data with a correlation coefficient of 0.419.

Therefore, the assumption made is valid. Hence, bigger and older ships are likely to have more miscellaneous group renewal weight than smaller and newer ships.

Table 8.1 displays the correlation coefficients of various best goodness of healthy relationships (linear trendline using average independent and dependent variables). It also shows the higher correlation coefficients against the (age * deadweight) variable. Therefore, variable (age * deadweight) may be considered more reliable than age and deadweight individually.

8.10 Regression

In the previous section and sub-sections, it has been highlighted that, theoretically, total structural steel renewal weight and its components (plate group, transverse

Table 8.1 Summary of correlation coefficients of various relationships of a linear trendline

Figure No.	Variables	r^2
8.1	R_S versus S_A	0.790
8.2	R_S versus S_D	0.688
8.4	R_S versus $(S_A * S_D)$	0.194
8.5	R_{PG} versus S_A	0.527
8.6	R_{PG} versus S_D	0.159
8.8	R_{PG} versus $(S_A * S_D)$	0.969
8.9	R_{TG} versus S_A	0.794
8.10	R_{TG} versus S_D	0.438
8.12	R_{TG} versus $(S_A * S_D)$	0.636
8.13	R_{LG} versus S_A	0.833
8.14	R_{LG} versus S_D	0.396
8.16	R_{LG} versus $(S_A * S_D)$	0.598
8.17	R_{MS} versus S_A	0.694
8.18	R_{MS} versus S_D	0.356
8.20	R_{MS} versus $(S_A * S_D)$	0.419

group, longitudinal group, and miscellaneous group) are directly and positively associated with age, deadweight, type and (age * deadweight) of the respective ship. In other words, structural steel renewal weight and its components independently (dependent variable) are a function of the age, deadweight, type and (age * deadweight) (independent variables) of a ship. Mathematically, the relationships mentioned above may be expressed in the equation form (Eqs. 8.1–8.20).

$$R_S = a + b * S_A \qquad (8.1)$$

$$R_S = a + b * S_D \qquad (8.2)$$

$$R_S = a + b * S_T \qquad (8.3)$$

$$R_S = a + b * (S_A * S_D) \qquad (8.4)$$

$$R_{PG} = a + b * S_A \qquad (8.5)$$

$$R_{PG} = a + b * S_D \qquad (8.6)$$

$$R_{PG} = a + b * S_T \qquad (8.7)$$

$$R_{PG} = a + b * (S_A * S_D) \tag{8.8}$$

$$R_{TG} = a + b * S_A \tag{8.9}$$

$$R_{TG} = a + b * S_D \tag{8.10}$$

$$R_{TG} = a + b * S_T \tag{8.11}$$

$$R_{TG} = a + b * (S_A * S_D) \tag{8.12}$$

$$R_{LG} = a + b * S_A \tag{8.13}$$

$$R_{LG} = a + b * S_D \tag{8.14}$$

$$R_{LG} = a + b * S_T \tag{8.15}$$

$$R_{LG} = a + b * (S_A * S_D) \tag{8.16}$$

$$R_{MS} = a + b * S_A \tag{8.17}$$

$$R_{MS} = a + b * S_D \tag{8.18}$$

$$R_{MS} = a + b * S_T \tag{8.19}$$

$$R_{MS} = a + b * (S_A * S_D) \tag{8.20}$$

Since all the independent variables are linearly associated with the dependent variable, a multiple linear regression model will likely fit the system. Accordingly, a multiple linear regression model is considered to establish the relationship between total structural steel renewal weight, age, deadweight, and type of ship. Components of total structural steel renewal weight are not considered for the regression equation. They will be dealt with in different approaches. The following function (Eq. 8.21) is chosen as per primarily mentioned assumptions.

$$R_S = f(S_A, S_D, S_T) \tag{8.21}$$

Table 8.2 Numerical values for types of ships for regression analysis

Types of ships	Numerical values for S_T for Eq. 8.21
Chemical Tanker	1.0000
Crude oil tanker	2.1183
General cargo carrier	2.1949
Bulk carrier	2.3420
Container carrier	7.7603
LPG carrier	9.1559

Table 8.3 Summary of values of statistical parameters of final regression equation

Statistical parameters	Values of statistical parameters for Equation 8.22
Sample size (n)	117
No. of independent variable (k)	3
Significance level (α)	0.05
Standard deviation (s)	9363
Coefficient of determination (R^2)	0.206
$f - f_\alpha$	7.05

Appropriate numerical values for S_T are calculated using average structural steel renewal weight per ship for each type and assigned for types for regression analysis. Table 8.2 display the numerical values assigned to the types for the functional equation (Eq. 8.21). It is used in regression analysis to form the regression Eq. 8.22.

Using the observed data for R_S, S_A, S_D, S_T and following the procedure of multiple regression analysis, the final regression equation (Eq. 8.22) is formed as follows,

$$R_S = -5178.852 + 624.327 * S_A - 0.00098 * S_D + 1145.940 * S_T \qquad (8.22)$$

The vital statistical parameters regarding the final regression Eq. (8.22) are given in Table 8.3. The condition $f > f_{0.05}$ (calculated F statistic and tabulated F statistic, respectively) suggested rejecting the null hypothesis. It may be concluded that there is a significant amount of variation in their response (the dependent variable) due to the differences in independent variables in the postulated models.

8.11 Validation

The validation technique is applied to Eq. 8.22. The summary of the validation results is presented in Table 8.4. The Table shows the outline of the variation of model values from the actual values. The deviation of model values from actual values explained in previous sections is equally applicable to structural steel renewal

Table 8.4 Summary of validation results of final regression equation

Items	Equation 8.22
Positive error (%)	45,586
Negative error (%)	−1465
Range of error (%)	47,051
Mean error (%)	1107
Variance	20,351,831
Standard deviation	4511

weight, too; only nature and magnitudes are different. Figure 8.21 demonstrates the relationship of deviation (%) in structural steel renewal weight against equivalent variation (%) in independent variables. The figure suggests that when the equivalent deviation (%) in independent variables is negative, the model will yield a lower value means negative variation (%) in structural steel renewal weight and vice versa. This behaviour entirely depends on the nature of the respective mathematical relationship (regression equations).

Validation is also applied to structural steel renewal weight relationships with age, deadweight, and (age * deadweight) (Figs. 8.1, 8.2 and 8.4), respectively. Results are presented in Table 8.5. It displays that validation results change significantly with

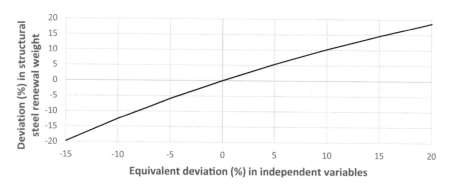

Fig. 8.21 Deviation in structural steel renewal weight versus equivalent deviation in independent variables

Table 8.5 Summary of validation results of structural steel renewal weight

Items	Figure 8.1	Figure 8.2	Figure 8.4
Positive error (%)	32,942	29,940	23,653
Negative error (%)	−117	−97	−96
Range of error (%)	33,059	30,037	23,749
Mean error (%)	943	1905	1218
Variance	11,717,536	22,904,812	10,899,742
Standard deviation	3423	4786	3301

Table 8.6 Summary of
validation results of plate
group renewal weight

Items	Figure 8.5	Figure 8.6	Figure 8.8
Positive error (%)	11,530	17,785	47,360
Negative error (%)	−92	−89	−73
Range of error (%)	11,622	17,873	47,433
Mean error (%)	481	1337	2566
Variance	2,204,574	7,786,096	40,704,202
Standard deviation	1485	2790	6380

the change of independent variable, especially for (age * deadweight). Furthermore, it indicates that (age * deadweight), as independent variables used in the analysis, significantly influence the structural steel renewal weight.

The same technique is applied to relationships of structural steel members' groups, such as plate group (Figs. 8.5, 8.6 and 8.8), transverse group (Figs. 8.9, 8.10 and 8.12), longitudinal group (Figs. 8.13, 8.14 and 8.16) and miscellaneous group (Figs. 8.17, 8.18 and 8.20) with age, deadweight and (age * deadweight). Results are presented in Tables 8.6, 8.7, 8.8 and 8.9. It displays that validation results for the relationship between structural group renewal weight and (age * deadweight) are better than age and deadweight. But for structural group renewal weight, it is different for different groups.

Finally, the proposed mathematical models may estimate the structural steel renewal weight and its components against various combinations of values for independent variables. Using the above model as a guide, shipyards may calculate the

Table 8.7 Summary of
validation results of
transverse group renewal
weight

Items	Figure 8.9	Figure 8.10	Figure 8.12
Positive error (%)	6000	3837	7425
Negative error (%)	−587	−96	−90
Range of error (%)	6587	3932	7514
Mean error (%)	616	325	809
Variance	1,530,099	477,028	1,865,043
Standard deviation	1237	691	1366

Table 8.8 Summary of
validation results of
longitudinal group renewal
weight

Items	Figure 8.13	Figure 8.14	Figure 8.16
Positive error (%)	12,621	23,460	13,228
Negative error (%)	−89	−98	−110
Range of error (%)	12,710	23,558	13,339
Mean error (%)	1665	1621	593
Variance	8,060,204	13,498,106	2,852,726
Standard deviation	2839	3674	1689

Table 8.9 Summary of validation results of miscellaneous group renewal weight

Items	Figure 8.17	Figure 8.18	Figure 8.20
Positive error (%)	11,122	13,615	20,344
Negative error (%)	−95	−98	−97
Range of error (%)	11,217	13,714	20,441
Mean error (%)	763	802	1159
Variance	3,475,435	3,290,147	7,098,561
Standard deviation	1864	1814	2664

expected scope of structural steel renewal works and the corresponding drydocking time, if necessary, including logistics. While using the model to estimate the scope of steel renewal works, one may be aware of the error level in predicting that. The model will provide a reasonable estimate of the expected independent variables that are close to the mean value. But for low and high value, the assessment will be low and high accordingly (Fig. 8.21). Also, one may consider allowing some allowance on top of the model value to accommodate various variations in structural steel renewal weight explained earlier.

8.12 General Conclusions

The present analysis reveals some basic but essential information regarding structural steel renewal weight. The highest quantity of structural steel renewal (average by type) was carried out in LPG carriers, and the lowest quantity of structural steel was carried out in chemical tankers irrespective of age and deadweight (Fig. 8.3).

Figure 8.22 displays the average structural steel renewal weight breakdown against the designated structural members like the plate, transverse, longitudinal, and miscellaneous groups. The highest contribution to complete structural steel renewal is from

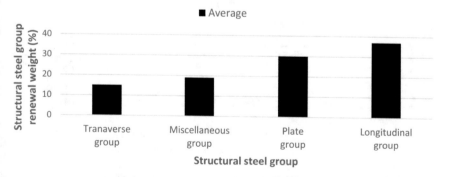

Fig. 8.22 Average structural steel renewal weight versus structural group

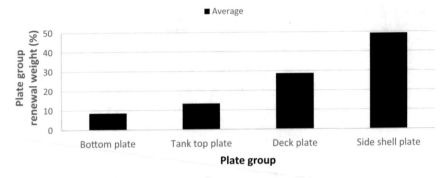

Fig. 8.23 Average plate group renewal weight versus plate group

the longitudinal group (37%), and the lowest contribution is from the transverse group (15%) irrespective of age and deadweight.

Figure 8.23 shows the components of the plate group and their contribution to the total plate group renewal by weight (%). The highest contribution recorded is from the side shell plate (49%), and the lowest contribution is from the bottom plate (9%).

Figure 8.24 shows the components of the transverse group and their contribution to the total transverse group renewal by weight (%). Again, the highest contribution is from side frames (72%), and the lowest contribution is from the bottom transverse (4%).

Figure 8.25 shows the component of the longitudinal group and their contribution to total longitudinal group renewal by weight (%). The highest contribution is from deck longitudinal (43%), and the lowest contribution is from longitudinal girders (1%).

Further analysis of the total structural steel renewal by the age group (with a span of five years) is presented in a graphical form in Fig. 8.26. It is evident from the figure that age has a significant influence on the renewal weight, but the responses

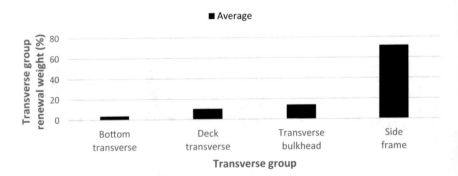

Fig. 8.24 Average transverse group renewal weight versus transverse group

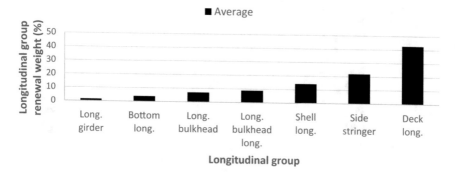

Fig. 8.25 Average longitudinal group renewal weight versus longitudinal group

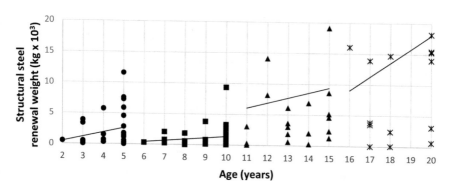

Fig. 8.26 Total structural steel renewal weight versus age group

are different for different age group spans. The higher age groups demand higher and quick answers compared to the lower age groups.

Further analysis of the total structural steel renewal weight by the deadweight group (with a span of 50,000 tonnes) is presented in a graphical form in Fig. 8.27. The figure reveals the inconsistent behaviour of the dependent variable with a wavy character and suggests that higher deadweight does not necessarily demand higher structural steel renewal.

The individual member of structural steel's contribution to the total steel renewal against ships' type is presented in both tabular and graphical form in Table 8.10 and Fig. 8.28, respectively. Similar data is presented in percentage form in both tabular and graphical form in Table 8.11 and Fig. 8.29, respectively. Thus, one can easily compare the average renewal weight of a particular structural member among the selected type of ships (Tables 8.10 and 8.11).

Figure 8.30 is developed using the relationships determined in Fig. 8.1 (linear trendline) between structural steel renewal weight and age. It displays the estimated structural steel renewal weight against age irrespective of deadweight and type.

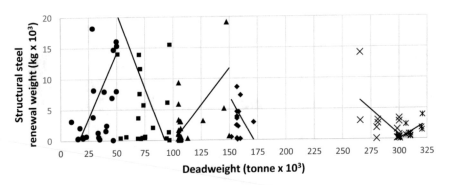

Fig. 8.27   Total structural steel renewal weight versus deadweight group

Table 8.10   Breakdown of structural steel renewal weight against type

Ships' type	Average steel renewal weight per ship (kg)				
	R_{PL}	R_{TR}	R_{LG}	R_{MS}	R_S
Crude Oil Tanker	1002	457	1008	678	3145
Container Carrier	3083	1344	4229	2792	11,448
Chemical Tanker	466	297	68	245	1076
Bulk Carrier	1572	1177	0	25	2774
LPG Carriers	3068	1718	7734	964	13,484

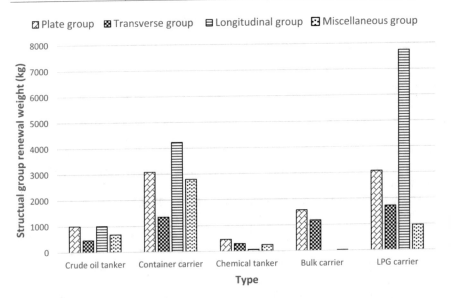

Fig. 8.28   Average structural group renewal weight versus type

Table 8.11 Breakdown of structural steel renewal weight against ship type

Ships' Type	Average steel renewal weight per ship (%)			
	R_{PG}	R_{TG}	R_{LG}	R_{MS}
Crude Oil Tanker	32	15	32	21
Container Carrier	27	12	37	24
Chemical Tanker	43	28	6	23
Bulk Carrier	57	42	0	1
LPG Carriers	23	13	57	7
Total	30	20	31	19

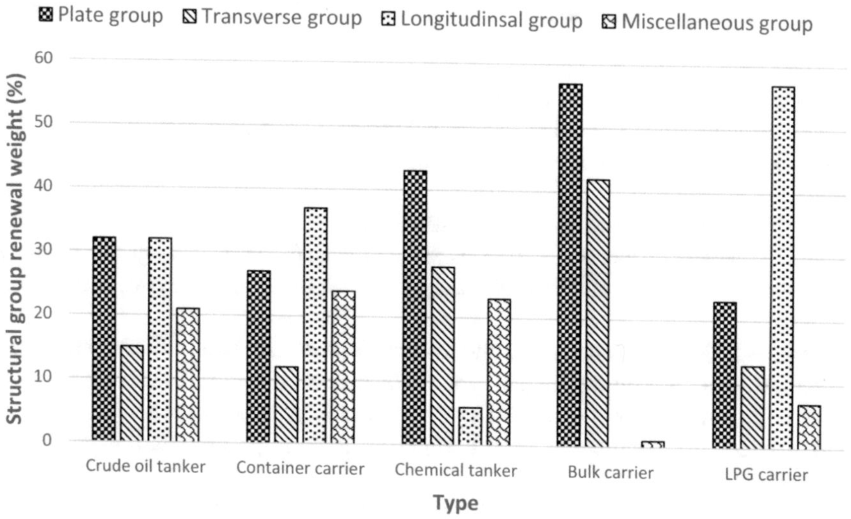

Fig. 8.29 Average structural group renewal weight versus type

Fig. 8.30 Estimated structural steel renewal weight versus age

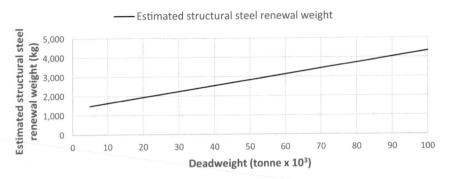

Fig. 8.31 Estimated structural steel renewal weight versus deadweight

Figure 8.31 is developed using the relationships determined in Fig. 8.2 (linear trendline) between structural steel renewal weight and deadweight. It displays the estimated structural steel renewal weight against deadweight irrespective of age and type.

Figure 8.32 is developed using the relationships determined in Fig. 8.4 (linear trendline) between structural steel renewal weight and (age * deadweight). It displays the estimated structural steel renewal weight against (age * deadweight) irrespective of type.

Figure 8.33 is developed using the relationships determined in Fig. 8.3 between structural steel renewal weight and type against age. It displays the estimated structural steel renewal weight against age for crude oil tankers, container carriers, bulk carriers, chemical tankers and liquified petroleum gas carriers irrespective of deadweight.

Figure 8.34 is developed using the relationships determined in Fig. 8.3 between structural steel renewal weight and type against deadweight. It displays the estimated structural steel renewal weight against deadweight for crude oil tankers,

Fig. 8.32 Estimated structural steel renewal weight versus (age * deadweight)

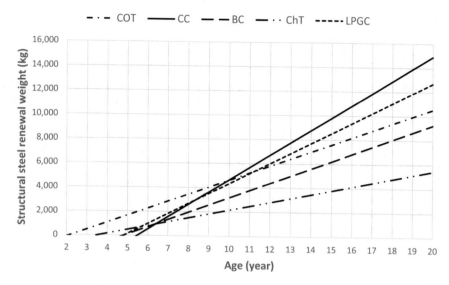

Fig. 8.33 Estimated structural steel renewal weight versus age for crude oil tankers, container carriers, bulk carriers, chemical tankers and liquified petroleum gas carriers

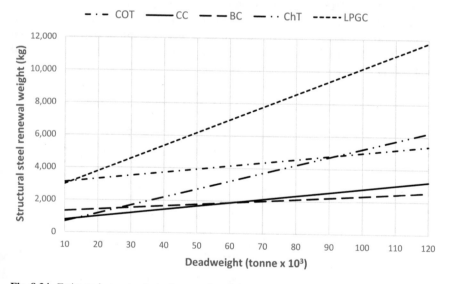

Fig. 8.34 Estimated structural steel renewal weight versus deadweight for crude oil tankers, container carriers, bulk carriers, chemical tankers and liquified petroleum gas carriers

container carriers, bulk carriers, chemical tankers and liquified petroleum gas carriers irrespective of deadweight.

Figure 8.35 is developed using the relationships determined in Fig. 8.3 between structural steel renewal weight and type against (age * deadweight). It displays the estimated structural steel renewal weight against (age * deadweight) for crude oil tankers, container carriers, bulk carriers, chemical tankers and liquified petroleum gas carriers.

Figure 8.36 is developed using the relationships determined in Fig. 8.22 between structural group renewal weight and structural group. It displays the estimated structural group renewal weight against the structural group.

Figure 8.37 is developed using the relationships determined in Fig. 8.23 plate group members, renewal weight and plate group renewal weight. It displays the estimated plate group members, renewal weight against plate group renewal weight.

Figure 8.38 is developed using the relationships determined in Fig. 8.24 transverse group members' renewal weight and transverse group renewal weight. It displays the estimated transverse group members, renewal weight against transverse group renewal weight.

Figure 8.39 is developed using the relationships in Fig. 8.25 longitudinal group members' renewal and renewal weights. It displays the estimated longitudinal group members, renewal weight against longitudinal group renewal weight.

Estimation of structural steel renewal weight from various viewpoints can be done using the respective appropriate relationship. One may also follow the options below with different independent variables to estimate structural steel renewal weight and components.

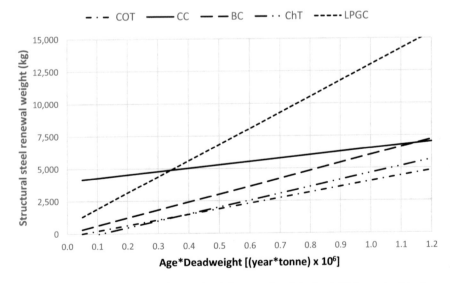

Fig. 8.35 Estimated structural steel renewal weight versus (age * deadweight) for crude oil tankers, container carriers, bulk carriers, chemical tankers and liquified petroleum gas carriers

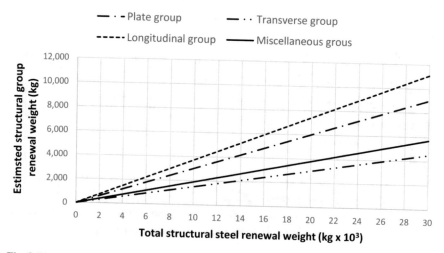

Fig. 8.36 Estimated structural group renewal weight versus total structural steel renewal weight

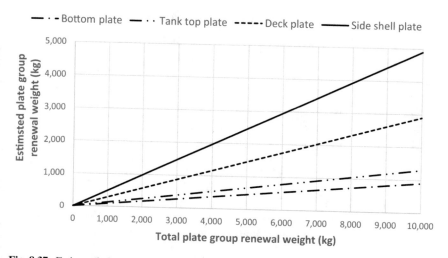

Fig. 8.37 Estimated plate group renewal weight versus total plate group renewal weight

Option—I

Use age and estimate the structural steel renewal weight irrespective of deadweight and type with the help of Fig. 8.30.

Option—II

Use deadweight and estimate the structural steel renewal weight irrespective of age and type with the help of Fig. 8.31.

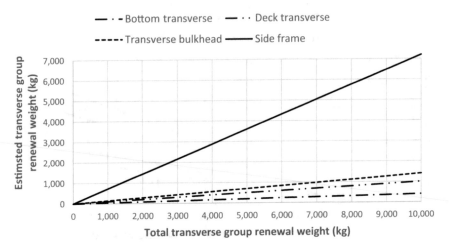

Fig. 8.38 Estimated transverse group renewal weight versus total transverse group renewal weight

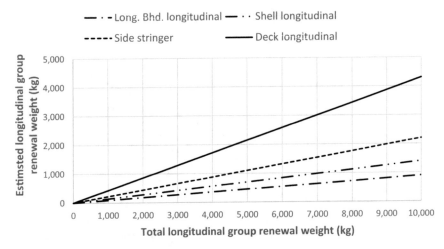

Fig. 8.39 Estimated longitudinal group renewal weight versus total longitudinal group renewal weight

Option—III

Use (age * deadweight) and estimate the structural steel renewal weight irrespective of type with the help of Fig. 8.32.

Option—IV

Use age and type and estimate the structural steel renewal weight for the corresponding type, irrespective of deadweight, with the help of Fig. 8.33 as appropriate.

Option—V

Use deadweight and type and estimate the structural steel renewal weight for the corresponding type, irrespective of age, with the help of Fig. 8.34 as appropriate.

Option—VI

Use (age * deadweight) and the type and estimate the structural steel renewal weight for the corresponding type with the help of Fig. 8.35 as appropriate.

Option—VII

Use structural steel renewal weight and estimate the structural group's renewal weight, irrespective of age, deadweight, and type, with the help of Fig. 8.36 as appropriate.

Option—VIII

Use plate group renewal weight and estimate the renewal weight of plate group members' renewal weight, irrespective of age, deadweight, and type with the help of Fig. 8.37 as appropriate.

Option—IX

Use transverse group renewal weight and estimate the renewal weight of transverse group members' renewal weight, irrespective of age, deadweight, and type with the help of Fig. 8.38 as appropriate.

Option X

Use longitudinal group renewal weight and estimate the renewal weight of longitudinal group members' renewal weight, irrespective of age, deadweight, and type with the help of Fig. 8.39 as appropriate.

Option—XI

Use age, deadweight and type and estimate the structural steel renewal weight for the corresponding type with the help of regression Eq. 8.22.

Finally, it is up to the individuals to use the findings according to their requirements and the variables available.

References

1. Dev, A. K., Saha, M.: Analysis of structural steel renewal weight in ship repairing. J. Ship Prod. Design **35**(2), 139–169 (2019)
2. Hiromi, S., Matsushita, H., Song, Y., Nakai, T., Yuya, N.: Thickness Reduction Due to Flow Accelerated Corrosion of Shipboard Piping. ClassNK Technical Bulletin, pp. 35–56 (2006)
3. Ozgur, U.S.: Interaction between ship repair, ship conversion and shipbuilding industries. Org. Econ. Co-oper. Dev. **2010**(3), 7–36 (2008)
4. Seref, A., Mirza, A.: Bulk carriers repair analysis. Ann. J. Soc. Naval Arch. Mar. Eng. **30**, 51–61 (2008). ISBN 978-981-4222-97-6
5. Yamamoto, N., Ikegami, K.: A study on the degradation of coating and corrosion of ship's hull based on the probabilistic approach. J. Offshore Mech. Arct. Eng. **120**, 121–128 (1998)

Chapter 9
Structural Steel Renewal Locations

9.1 Introduction

In Chap. 8, the focus is on quantifying the weight of structural steel renewal from various viewpoints. In this Chapter, the focus will be on identifying locations of structural steel renewal from multiple perspectives. This information will help the shipowners identify the expected structural steel renewal locations and prepare the ship accordingly before proceeding to the repair yard.

Structural steel renewal in ship repairing is routine work throughout the entire service/operational life. Prior information about the quantity of structural steel to be renewed helps shipowners allocate an appropriate budget, and the shipyard can prepare a proper ship repairing schedule. Similarly, preliminary information about the renewal of structural members and their renewal locations can help the shipowner prepare the ship before going to the yard to reduce the loss time. In addition, the shipyard can plan for appropriate logistics. For example, if the renewal locations of side shell plates are in the way of a fuel oil tank in the engine room, then the respective tank(s) can be made empty for cleaning. Following this, the commencement of cleaning the tank can occur before arrival at the yard.

Similarly, suppose this information is made available to the shipyard. In that case, they can assess and decide if the steel renewal works can be done afloat or in the drydock or both, and accordingly, the shipyard can prepare the repairing schedule, including dock-in/dock-out time (no. of days in drydock).

Being a floating structure, a ship is a natural complex structure compared to any other land-based steel structure. Its structural complexity is due to the environmental conditions in which a ship operates throughout her life and governs by the rules and

© The Author(s), under exclusive license to Springer Nature Singapore Pte Ltd. 2022
A. K. Dev et al., *Ship Repairing*, Springer Series on Naval Architecture,
Marine Engineering, Shipbuilding and Shipping 12,
https://doi.org/10.1007/978-981-16-9468-4_9

regulations of classification societies. However, for simplicity and ease to understand, the following primary steel structural members are selected for analysis in this chapter.

Plates

1. Deck plates (including deck stringers)
2. Side shell plates (including sheer strakes and bilge strakes)
3. Bottom shell plates (including keel plate and garboard strakes).

Transverse Members

1. Deck transverses
2. Side frames: Vertical transverse webs on side shell plates (commonly known as web frames)
3. Side frames: Vertical transverse webs on longitudinal bulkheads
4. Bottom transverses
5. Transverse bulkheads including stiffeners.

Longitudinal Members

1. Deck longitudinals
2. Side shell longitudinals
3. Bottom shell longitudinals
4. Longitudinal bulkheads
5. Longitudinal bulkhead longitudinals.

As per the current practice, the structural steel renewal is known only when renewal work is confirmed after a successful joint inspection of the hull (external and internal). Accordingly, ship crews commence preparing tanks/compartments involved, if required. For example, if any fuel oil tank/lubricating oil tank is concerned, they transfer the oil to other tanks. Then the yard starts to mobilize the necessary logistics to clean the tank for hot work certification. It causes a loss of time, which increases the number of days for a ship in the yard, hence the cost for shipowners. It also causes loss of earnings for the owner and loss of revenue for the yard.

This situation can be improved significantly if prior information about structural steel renewal locations is known to both the shipowners and the shipyards. This improvement can easily be transformed into a saving of days which saves cost for the shipowners and increases revenue for shipyards [1].

In the analysis, attention is paid to highlight, investigate, and determine the interrelationship among the dependent (renewal locations) and the independent variables (principal dimensions of appropriate combination, age, and length) responsible for renewal locations of structural steel.

Ships with routine drydocking repair are considered only, not emergency repair and damage repair. As mentioned above, the purpose is to get a more uniform (stable), reliable, and realistic relationship among the variables. Therefore, analysis of collected data (renewal locations) is carried out against the principal dimensions (as appropriate), age and length of ships.

There is no documented information available about structural steel renewal locations in ship repairing concerning their age, size (deadweight), and type. No academic paper, article or research works have been found devoted to structural steel renewal locations of ships regarding their age, deadweight, and type. The probable reasons seem to be the scarcity and confidentiality of such classified information and data. However, some works, not exactly but close to the issue, were done from different viewpoints. Seref and Mirza [6] reviewed structural steel renewal works of bulk carriers. Through the analysis, areas prone to structural deterioration are highlighted. The paper demonstrates the locations and trends of repairs caused by factors related to ships' operations, exposure to natural elements of their operating environment at sea, and quality of ship staff. All these point to a predictive pattern where repairs are most likely to occur in bulk carriers. Dev and Saha [1] evaluated structural steel renewal locations concerning the ship's principal dimensions. The results of various analyses of renewal locations of multiple members confirm that the structural elements, primary or secondary, in two corners (intersection of the deck and side shell plate—port and starboard), suffer some sorts of failure, which results in steel replacement irrespective of age and length and any other parameters of a ship. These areas are prone to be affected by the ship's movement and motion, such as hogging, sagging, heaving, pitching, and rolling, and required to be inspected/surveyed very carefully and with particular attention to avoid any severe consequences. A conclusion was drawn that the length and age of a ship do not dictate the renewal locations of structural members. ISSC [3] has elaborated about the inspection and monitoring of the structure's health using scientifically advanced techniques that help prevent the structure's sudden failure. However, it did not elaborate on the locations of inspection and monitoring. Smith [7] demonstrated the evidence of extreme waves, their significant wave heights, and their impacts on the structural design of the ships. Also highlighted, the significant wave heights are increased from 11 to 20 m, and at the same time, significant wave heights of 30 m are not ruled out. It is suggested that the dynamic force of the wave impacts should also be included in the structural analysis of the ship, including hatch covers and other vulnerable areas. It is concluded that a design criterion based on an 11 m high significant wave seems inadequate and should be at least 20 m.

9.2 Structural Steel Renewal Locations Versus Principal Dimensions

This analysis focuses on the actual renewal locations of different structural members concerning the ship's dimensions (longitudinal direction, transverse direction, and vertical direction). From an engineering point of view, it also identifies the possible reason, explains/justifies the renewal locations of selected structural members. To compare the steel renewal locations of all sample ships (those are all different dimensions), the actual locations of structural steel renewal of each ship are converted in terms of fraction (%) of respective ship's dimensions (longitudinally x-axis, transversely y-axis and vertically z-axis). These are the length between perpendicular, breadth and depth, as required and appropriate.

For the analysis, selected structural members are chosen based on the frequency (number of cases) of renewal, such as deck plates, side shell plates, bottom shell plates, deck longitudinals, side shell longitudinals, bottom shell longitudinals, longitudinal bulkhead longitudinals, deck transverses, side frames (vertical transverse webs on side shell plates and longitudinal bulkheads) and side stringers.

The renewal locations of structural steel members were identified from the repair plans collected from the shipyard, and the relative positions were determined concerning the ship's centreline (longitudinal), midship transverse line and baseline for each ship. These locations were then converted to the fraction (%) of the ship's dimensions (length, breadth, and depth) to get the coordinates concerning the longitudinal centreline (as the x-axis), transverse centreline (as the y-axis) and vertical line (z-axis). Finally, the converted points (coordinates) are plotted and presented in the figures.

9.2.1 Deck Plates

The renewal locations of deck plates are presented in Figs. 9.1, 9.2 and 9.3 from different viewpoints. Figure 9.1 shows the renewal locations to the length (%) and breadth (%) of a ship and suggests that most renewals were carried out at the transverse end of deck plates (port and starboard). Figure 9.2 demonstrates the frequency (%) of renewal locations of deck plates regarding the longitudinal centreline and breadth of a ship. Those also show that transversely, most of the renewal locations of deck plates (about 62%) were located within 20% of half breadth (10% of breadth) from side shell plates (port and starboard). Figure 9.3 presents the frequency (%) of renewal locations of deck plates about the length. Those also show that longitudinally, most of the renewals of deck plates (about 65%) were located at the forward

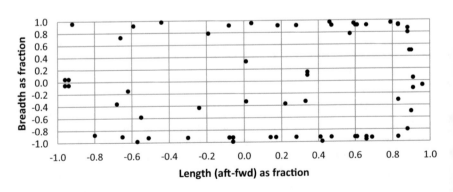

Fig. 9.1 Renewal locations of deck plates

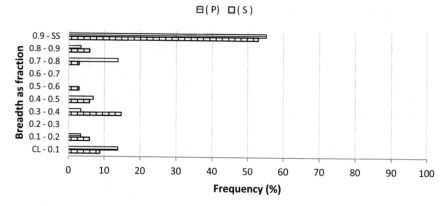

Fig. 9.2 Distribution of renewal locations of deck plates along half breadth (P and S)

of the midship and extended up to the forecastle deck. The possible reasons for the renewal locations of deck plates are probably the ship's movement in rough weather under which a ship usually experiences oscillatory, translatory and torsional motion.

Deck plates are one of the prominent structural members of a ship. In real-life situations, deck plates, alternately, experience tensile and compressive loads together with torsional effects due to the ship's motions, like hogging, sagging, rolling, pitching, heaving, etc. Deck plates, together with stiffening members (transverse and longitudinal), significantly contribute to the longitudinal strength like a considerable box-girder.

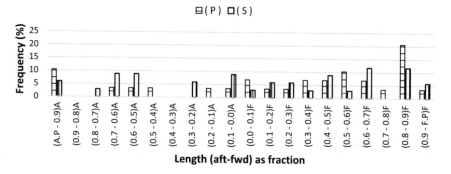

Fig. 9.3 Distribution of renewal locations of deck plates along length (P and S)

9.2.2 Side Shell Plates

From different viewpoints, the renewal locations of side shell plates (port and star-board) are presented in Figs. 9.4, 9.5 and 9.6. Figure 9.4 highlights the renewal locations of side shell plates for length (%) and depth (%) of a ship at port and starboard, respectively. Figure 9.5 displays the frequency (%) of renewal locations of side shell plates regarding the height from the baseline to the main deck level. Those also show that vertically, about 55% of renewals lie within 40% of depth from the deck. Figure 9.6 depicts the frequency (%) of renewal locations of side shell plates concerning length. Those also show that longitudinally, a significant portion of renewal locations (about 58%) is located within 20% of the length from forward perpendicular and aft perpendicular and about 37% in the way of midship area. The distribution pattern of renewal locations suggests that the impact of push from a tugboat during mooring and un-mooring, impact of wave forces on the forward

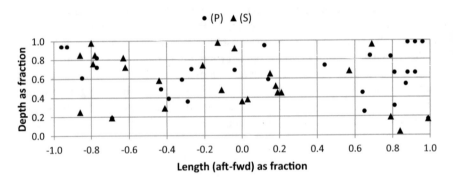

Fig. 9.4 Renewal locations of side shell plates (P and S)

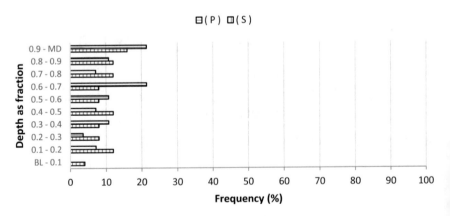

Fig. 9.5 Distribution of renewal locations of side shell plates along depth (P and S)

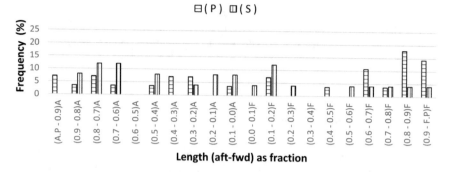

Fig. 9.6 Distribution of renewal locations of side shell plates along length (P and S)

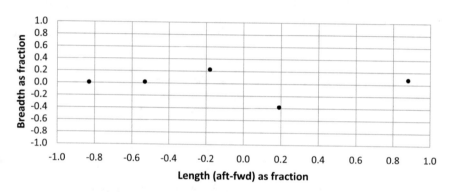

Fig. 9.7   Renewal locations of bottom shell plates along length

and aft, the effect of panting and racking are the main reasons. However, the ship's movement also plays a part.

Side shell plates are one of the prominent structural members of a ship. Together with stiffening members (transverse and longitudinal), it contributes to the ship's longitudinal strength. It also withstands hydrostatic pressure and wave forces. Hydrostatic pressure and wave load vary with the ship's motions.

9.2.3   Bottom Shell Plates

Renewal of bottom shell plates is very rare. Probably, it is well protected, externally, against corrosion and fouling and internally against corrosion. Usually, the bottom shell plate is renewed if the vessel is grounded and suffers damages, or it runs over underwater objects and suffers damages. Five cases of bottom plate renewal are recorded (Fig. 9.7) and are not enough to draw a fair conclusion. As such, it is omitted from the analysis. However, bottom shell plates are one of the prominent structural members of a ship. It also withstands hydrostatic pressure and vertical wave forces. Hydrostatic pressure and wave load vary with the ship's motions. Bottom shell plates and stiffening members, transverse and longitudinal, significantly contribute to a ship's longitudinal strength like a considerable box-girder.

9.2.4 Deck Longitudinals

The renewal of deck longitudinals is presented in Figs. 9.8, 9.9 and 9.10 from different viewpoints. Figure 9.8 introduces renewal locations of deck longitudinals concerning the length (%) and breadth (%) of a ship. Figure 9.9 produces the frequency (%) of renewal of deck longitudinals regarding longitudinal central line and breadth of the ship. Those also show that transversely, most of the renewal of deck longitudinals (about 91%) are located within 20% of half breadth (10% of breadth) from the side

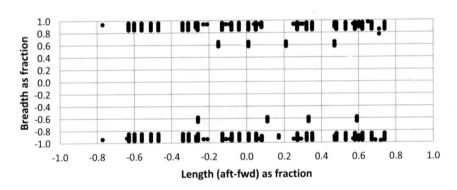

Fig. 9.8 Renewal locations of deck longitudinals along length

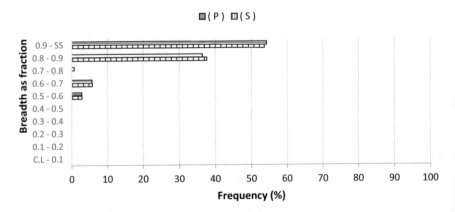

Fig. 9.9 Distribution of renewal locations of deck longitudinals along half breadth (P and S)

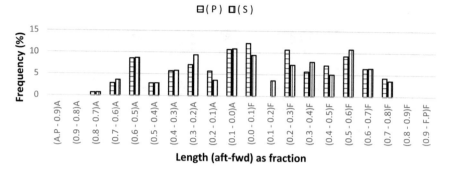

Fig. 9.10 Distribution of renewal locations of deck longitudinals along length (P and S)

shell plate, in both port and starboard. Figure 9.10 shows the frequency (%) of the renewal of deck longitudinals about length. Those also show that almost all the renewals are located within 35% of length, forward and aft from midship longitudinally. Renewal locations concerning breadth strongly suggest ship's movement, mainly rolling, pitching, hogging, and sagging, are the root cause of these locations. Moreover, locations are in the wing tank (usually ballast tank), so the tank environment also plays a role.

Deck longitudinals are attached to the deck plates to stiffen them longitudinally and increase the buckling strength of the plating buckling strength. In addition, since these longitudinals are effectively connected to the plating, those contribute to the general longitudinal strength of the structure. Hence, deck longitudinals may be considered secondary structural members.

9.2.5 Side Shell Longitudinals

The renewal locations of side shell longitudinals (port and starboard) are presented in Figs. 9.11, 9.12 and 9.13 from different viewpoints. Figure 9.11 exhibits the renewal locations of side shell longitudinals concerning the length (%) and depth (%) of a ship at port and starboard. Figure 9.12 demonstrates the frequency (%) of renewal of side shell longitudinals regarding the height from the baseline to the main deck of a ship. Those also show that about 70% of renewal vertically lies within 20% of depth from the main deck (port and starboard). Figure 9.13 presents the frequency (%) of side shell longitudinals renewal locations concerning length. Those also show that most renewal (about 67%) lies in the forward of midship longitudinally. A significant amount of renewal was also carried out about the midship area. Longitudinally, renewal locations are almost identical to that of deck longitudinals (Fig. 9.8). Vertical locations from the deck or baseline suggest the hogging, sagging, and twisting is the root causes. Effects of tank environments are not ruled out too.

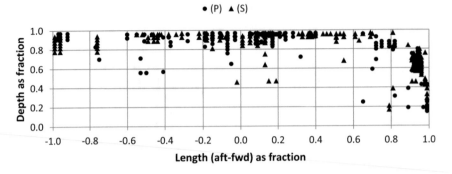

Fig. 9.11 Renewal locations of side shell longitudinals along length (P and S)

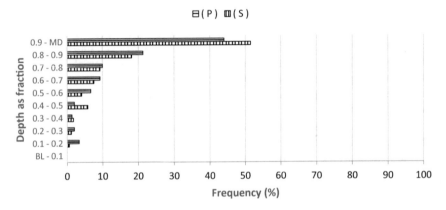

Fig. 9.12 Distribution of renewal locations of side shell longitudinals along depth (P and S)

Side shell longitudinals, like deck longitudinals, are attached to the side shell plate to stiffen it longitudinally and increase the plating buckling strength. In addition, since

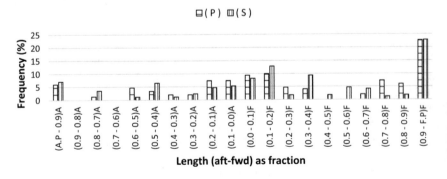

Fig. 9.13 Distribution of renewal locations of side shell longitudinals along length (P and S)

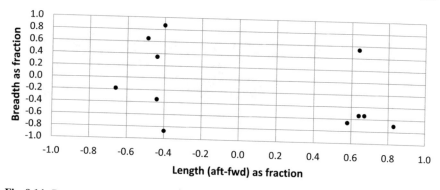

Fig. 9.14 Renewal locations of bottom shell longitudinals along length

these longitudinals are effectively connected to the plating, those contribute to the general longitudinal strength of the structure. Hence, side shell longitudinals may be considered secondary structural members.

9.2.6 Bottom Shell Longitudinals

Like bottom shell plates, the renewal of bottom shell longitudinals (port and starboard) is always very rare. Only eleven cases of renewal cases of bottom shell longitudinals were recorded (Fig. 9.14) and not enough to draw a fair conclusion. As such, it is omitted from the analysis. However, bottom shell longitudinals, like deck longitudinals, are attached to the bottom shell plate to stiffen the bottom shell plate longitudinally and increase the buckling strength of the plating. Since these longitudinals are effectively connected to the plating, those contribute to the general longitudinal strength of the structure. Hence, bottom shell longitudinals may be considered secondary structural members.

9.2.7 Longitudinal Bulkhead Longitudinals

The renewal locations of longitudinal bulkhead longitudinals (port and starboard) are presented in Figs. 9.15, 9.16 and 9.17 from different viewpoints. Figure 9.15 highlights the renewal locations of longitudinal bulkhead longitudinals concerning length (%) and depth (%) at the port and starboard. Figure 9.16 displays the frequency (%) of renewal of longitudinal bulkhead longitudinals regarding the height from baseline to the main deck of a ship. Those also show that vertically, almost 57% of renewals are within 20% of depth from the main deck (port and starboard). Figure 9.17 depicts the frequency (%) of renewal locations of longitudinal bulkhead longitudinals about length. It also shows that most renewal (60%) lies in the forward of midship

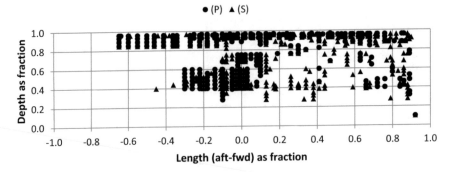

Fig. 9.15 Renewal locations of longitudinal bulkhead longitudinals along length (P and S)

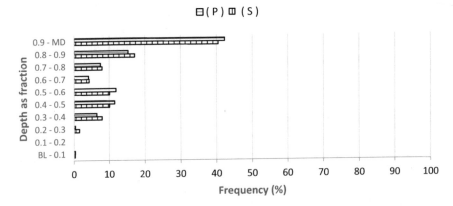

Fig. 9.16 Distribution of renewal locations of longitudinal bulkhead longitudinals along depth (P and S)

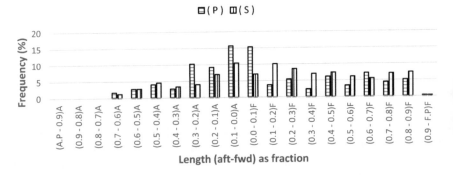

Fig. 9.17 Distribution of renewal locations of longitudinal bulkhead longitudinals along length (P and S)

longitudinally. Also, about 60% of renewal lies within 40% of half-length, in both forward and aft of midship. Longitudinally and vertically, renewal locations are identical to that of side shell longitudinals (Fig. 9.11) and the result of hogging, sagging, and twisting together with wing tank environments.

The longitudinal bulkhead is one of the primary structural members of a ship. It provides longitudinal strength to the structure. Longitudinals are attached to the bulkhead to stiffen the bulkhead plates and increase the buckling strength of the plating. Since these longitudinals are effectively connected to the plating, those contribute to the general longitudinal strength of the structure. Hence, longitudinal bulkhead longitudinals may be considered secondary structural members.

The behaviour of renewal locations of longitudinal bulkhead longitudinals (Fig. 9.15) and side shell longitudinals (Fig. 9.11) are very much similar, both vertically and longitudinally.

9.2.8 Deck Transverses

The renewal locations of deck transverses are presented in Figs. 9.18, 9.19 and 9.20 from different viewpoints. Figure 9.18 introduces the renewal locations of deck transverses concerning the length (%) and breadth (%) of a ship. Figure 9.19 produces the frequency (%) of renewal of deck transverses regarding the longitudinal central line and breadth of the ship. Those also show that transversely, most of the renewal of deck transverses (about 90%) are located within 20% of half breadth (10% of breadth) from the side shell plates, in both port and starboard. Figure 9.20 shows the frequency (%) of the renewal of deck transverses regarding length. Those also show that longitudinally, about 55% of renewals are in the forward of midship. Also, about 61% of renewals are within 40% of half-length from midship forward and aft. Longitudinally and transversely, renewal locations are identical to deck longitudinals (Fig. 9.8), and they are in the same compartment (wing tank). Their root causes are also similar.

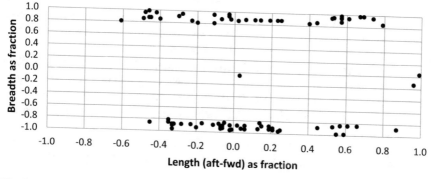

Fig. 9.18 Renewal locations of deck transverses along length

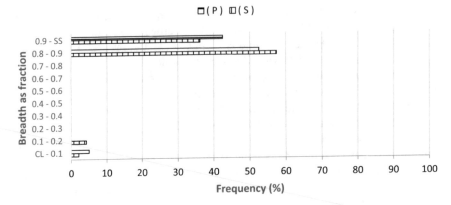

Fig. 9.19 Distribution of renewal locations of deck transverses along half breadth (P and S)

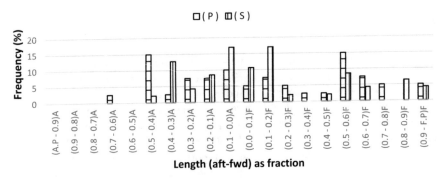

Fig. 9.20 Distribution of renewal locations of deck transverses along length (P and S)

A ship can be considered a massive box-girder of critical dimensions related to its shell plate thickness. Therefore, unless the shell plating is stiffened correctly in some way throughout the length of the ship, it will be incapable of withstanding compressive loads. For this reason, the shell plating is stiffened longitudinally and transversely. This transverse ring stiffening consists of three parts: in the double bottom called floor plates (bottom transverses), at the shipside called side frames (web frames) and under the deck called deck transverses.

So, the deck transverse is a part of transverse framing. It is fitted to the deck plate (generally underside) to stiffen the deck plate transversely and increase the buckling strength of the plating transversely. This transverse with its associated plating forms an effective built-up girder. These deck transverses may be considered as the primary structural members.

The pattern of renewal locations of deck transverses (Fig. 9.18) and deck longitudinals (Fig. 9.8) are very similar, both transversely and longitudinally.

9.2.9 Side Frames (Vertical Transverse Webs on Side Shell Plates)

The renewal locations of vertical transverse webs on side shell plates (port and starboard) are presented in Figs. 9.21, 9.22 and 9.23 from different viewpoints. Figure 9.21 exhibits the renewal locations of vertical transverse webs on the side shell plate concerning the length (%) and depth (%) of a ship at port and starboard. Figure 9.22 demonstrates the frequency (%) of renewal of side frames (vertical transverse webs on side shell) regarding height from the baseline to the main deck of a ship. Those also show that vertically, about 70% of the renewals are within 40% of depth from the main deck (both port and starboard). Figure 9.23 presents the frequency (%) of renewal locations of side frames (vertical transverse webs on side shell) about length. Those also show that longitudinally, about 60% of renewals are in the forward of midship with a concentration of 44%. Those renewal locations are

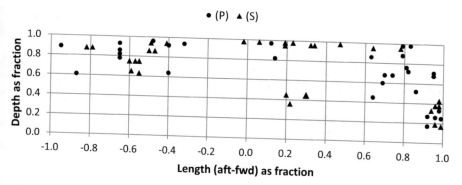

Fig. 9.21 Renewal locations of side frames i.w.o. side shell plates along length (P and S)

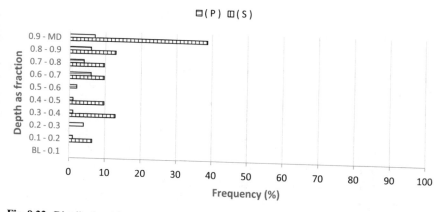

Fig. 9.22 Distribution of renewal locations of side frames i.w.o. side shell plates along depth (P and S)

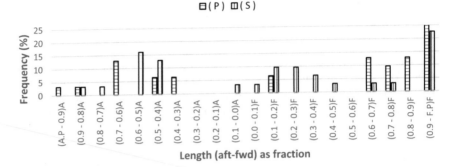

Fig. 9.23 Distribution of renewal locations of side frames i.w.o. side shell plates along length (P and S)

within 40% of half-length from forward perpendicular. Vertically, renewal locations are like side shell longitudinals (Fig. 9.11) and in the deck and side shell intersection area. Ship's movement like hogging, sagging, and twisting are the possible root causes, including tank environment.

As mentioned earlier under Sect. 9.2.8, the vertical transverse web on the side shell plate is a part of the transverse framing (ring form-all around the transverse section) and fitted in between side shell plates and longitudinal bulkheads. As mentioned earlier, the side shell plate is constantly subjected to hydrostatic pressure, wave forces, etc. Together with side shell longitudinals, these side frames stiffen the side shell plate longitudinally and increase the plating's buckling strength and distortion strength. These side frames are primarily to increase the transverse strength of the hull. However, side frames are one of the prominent structural members of the ship's hull girder structure.

9.2.10 Side Frames (Vertical Transverse Webs on Longitudinal Bulkhead)

The renewal locations of vertical transverse webs on longitudinal bulkheads (port and starboard) are presented in Figs. 9.24, 9.25 and 9.26 from different viewpoints. Figure 9.24 exhibits the renewal locations of vertical transverse webs on longitudinal bulkheads concerning the length (%) and depth (%) at the port and starboard. Figure 9.25 demonstrates the frequency (%) of renewal of side frames (vertical transverse webs on longitudinal bulkhead) regarding the height from the baseline to the main deck of a ship. Those also show that vertically, about 66% of renewals are within 20% of depth from the main deck (both port and starboard). Figure 9.26 highlights the frequency (%) of renewal locations of side frames (vertical transverse webs on longitudinal bulkhead) about length. Those also show that longitudinally, about 61% of the renewals are in the forward of midship. Also, about 45% of renewals are

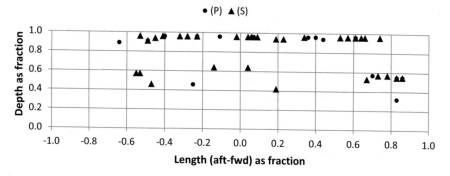

Fig. 9.24 Renewal locations of side frames i.w.o. longitudinal bulkhead along length (P and S)

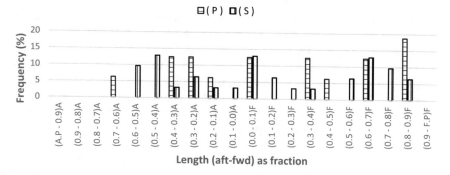

Fig. 9.25 Distribution of renewal locations of side frames i.w.o. longitudinal bulkhead along depth (P and S)

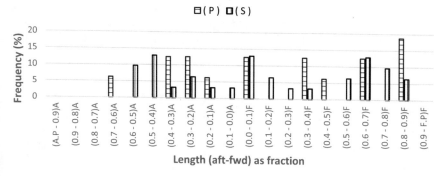

Fig. 9.26 Distribution of renewal locations of side frames i.w.o. longitudinal bulkhead along length (P and S)

within 40% of half-length from midship, in both forward and aft. Vertically, renewal locations are like side frames in the way of side shell (Fig. 9.21). The possible root causes are identical, too, including the tank environment.

The pattern of renewal locations of side frames (vertical transverse webs on longitudinal bulkheads) (Fig. 9.24) and side shell plates (Fig. 9.21) are very much similar, both vertically and longitudinally.

9.2.11 Side Stringers

The renewal locations of side stringers (port and starboard) are displayed in Fig. 9.27. It shows the renewal locations of side stringers concerning length (%) and breadth (%) of a ship at port and starboard. This Figure is identical to the renewal locations of deck longitudinals (Fig. 9.8) and deck transverses (Fig. 9.18) by nature. As far as side stringer renewal locations are concerned, the transverse position from the centreline does not matter. It is guided by the width of the wing tank of a ship but the longitudinal position of renewal matters. It is significantly influenced by the ship's movement, like side shell longitudinals. Figure 9.27 shows that the renewals of side stringer (port and starboard) were carried out up to 70% of the half-length forward of midship and 60% of the half-length aft of midship. A cluster of renewal locations is recorded at the forward (beyond 90% half-length). These locations are flat and stringer in the forepeak tank at the bosun store and possibly due to panting effects.

Side stringers are fitted longitudinally in the wing tank connecting the side shell plate and the longitudinal bulkhead plate. Under its position stiffens side shell plates and longitudinal bulkhead plates and increases the buckling strength of side shell plates and longitudinal bulkhead plates. Its contribution to the ship's structure is like that of side shell longitudinals and longitudinal bulkhead longitudinals but with greater magnitude. Side stringer may be considered as a central structural member of the ship's structure.

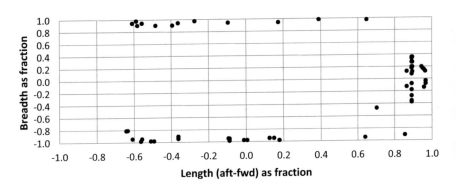

Fig. 9.27 Renewal locations of side stringers along length (P and S)

9.3 Structural Steel Renewal Locations Versus Age

This analysis investigates the renewal locations of selected structural members such as deck plates, deck longitudinals, deck transverses, side shell plates, side shell longitudinals, longitudinal bulkhead longitudinals, vertical transverse webs on side shell plates and vertical transverse webs on longitudinal bulkheads (in line with Sect. 9.2) concerning the age of a ship. The findings are presented in Figs. 9.28, 9.29, 9.30, 9.31, 9.32, 9.33, 9.34, 9.35, 9.36, 9.37, 9.38, 9.39, 9.40 and 9.41. This analysis aims to examine and determine the effects of the age of a ship on the renewal locations of the previously mentioned steel structural members.

A close look to Figs. 9.28, 9.29, 9.30, 9.31, 9.32, 9.33, 9.34, 9.35, 9.36, 9.37, 9.38, 9.39 and 9.40 reveals that the age of a ship does not directly influence the renewal locations of mentioned structural members. Whether an old or a new ship, the pattern of renewal locations concerning ship dimensions remains the same. The probable reason is that the renewal locations are greatly influenced by the movements of ships, as explained in Sect. 9.2 under sub-sections for various structural members. However, the analysis also reveals that the age of those ships were five years and

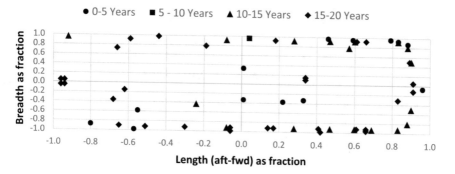

Fig. 9.28 Renewal locations of deck plates along length

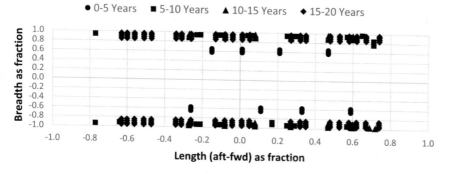

Fig. 9.29 Renewal locations of deck longitudinals along length

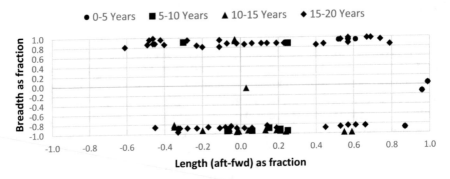

Fig. 9.30 Renewal locations of deck transverses along length

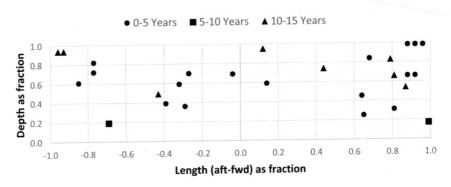

Fig. 9.31 Renewal locations of side shell plates along length (P)

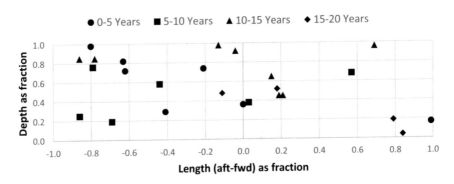

Fig. 9.32 Renewal locations of side shell plates along length (S)

above. More specifically, vessels of age groups 10–15 and 15–20 suffer the highest combined number of renewal locations irrespective of length, deadweight, and type, as demonstrated in Fig. 9.41. This finding agrees with the finding in Chap. 8 regarding

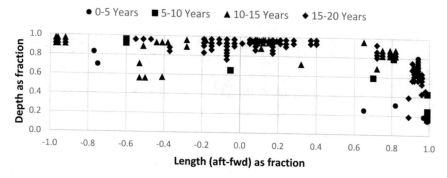

Fig. 9.33 Renewal locations of side shell longitudinals along length (P)

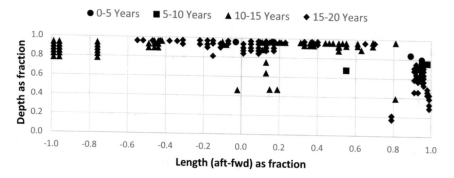

Fig. 9.34 Renewal locations of side shell longitudinals along length (S)

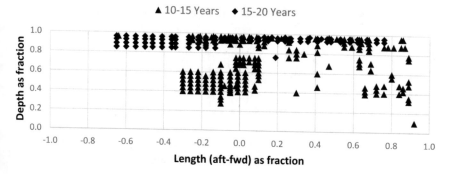

Fig. 9.35 Renewal locations of longitudinal bulkhead longitudinals along length (P)

structural steel renewal weight that older ships will see more structural steel renewal than new ships.

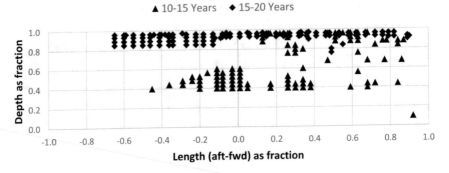

Fig. 9.36 Renewal locations of longitudinal bulkhead longitudinals along length (S)

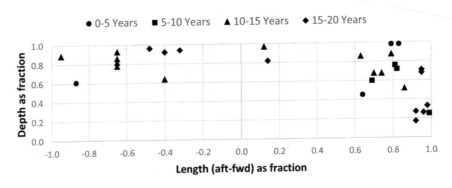

Fig. 9.37 Renewal locations of side frames i.w.o. side shell plates along length (P)

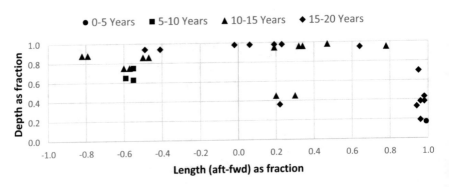

Fig. 9.38 Renewal locations of side frames i.w.o. side shell plates along length (S)

9.4 Structural Steel Renewal Locations Versus Length

This analysis relooks on the renewal locations of selected structural members such as deck plates, deck longitudinals, deck transverses, side shell plates, side shell

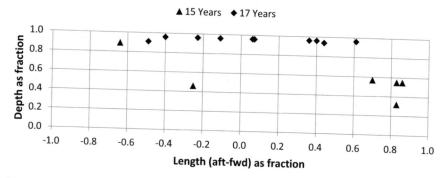

Fig. 9.39 Renewal locations of side frames i.w.o. longitudinal bulkhead along length (P)

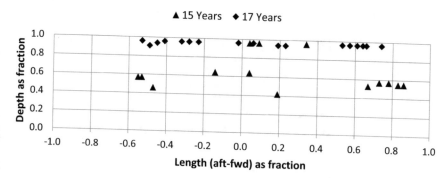

Fig. 9.40 Renewal locations of side frames i.w.o. longitudinal bulkhead along length (S)

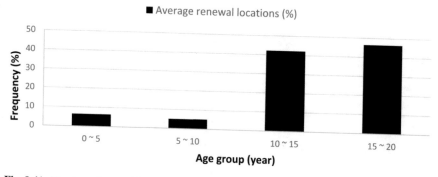

Fig. 9.41 Number of renewal locations versus age group

longitudinals, longitudinal bulkhead longitudinals, vertical transverse webs on side shell plates and vertical transverse webs on longitudinal bulkheads (in line with Sect. 9.3) concerning the length of a ship and the results are presented in Figs. 9.42, 9.43, 9.44, 9.45, 9.46, 9.47, 9.48, 9.49, 9.50, 9.51, 9.52, 9.53 and 9.54. Thus, this

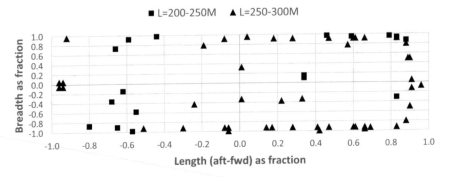

Fig. 9.42 Renewal locations of deck plates along length

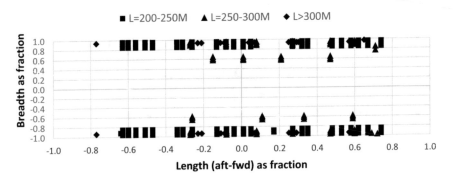

Fig. 9.43 Renewal locations of deck longitudinals along length

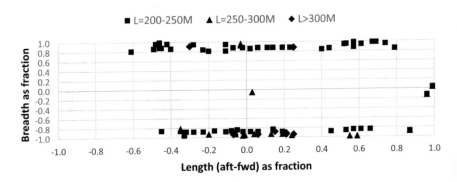

Fig. 9.44 Renewal locations of deck transverses along length

analysis aims to examine and determine the effect of the length of a ship on the renewal locations of the previously mentioned steel structural members.

Like age, the ship's length does not directly influence the renewal locations of the above-mentioned structural members. Whether a bigger or a smaller ship, the pattern

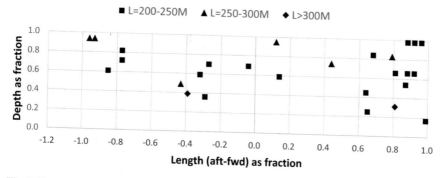

Fig. 9.45 Renewal locations of side shell plates along length (P)

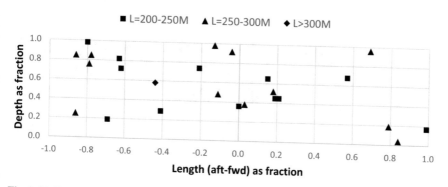

Fig. 9.46 Renewal locations of side shell plates along length (S)

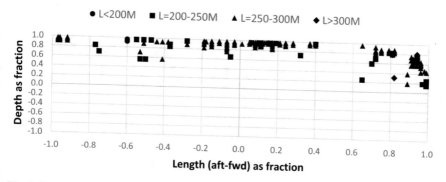

Fig. 9.47 Renewal locations of side shell longitudinals along length (P)

of renewal locations concerning ship dimensions remains the same. The probable reason is that the renewal locations are greatly influenced by the movements of ships, as explained in Sect. 9.2 under sub-sections for various structural members. However, the analysis also reveals that the ships in question were more than 200 m in

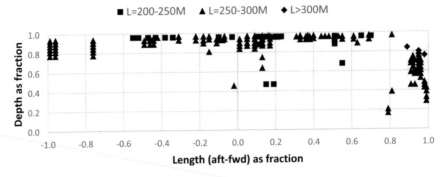

Fig. 9.48 Renewal locations of side shell longitudinals along length (S)

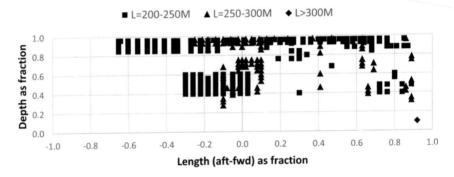

Fig. 9.49 Renewal locations of longitudinal bulkhead longitudinals along length (P)

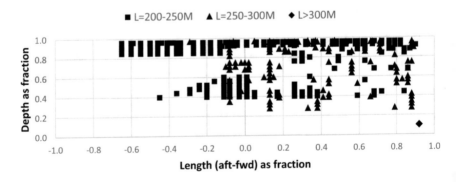

Fig. 9.50 Renewal locations of longitudinal bulkhead longitudinals along length (S)

length. More specifically, vessels of length groups 200–250 m and 250–300 m suffer the highest combined number of renewal locations irrespective of age and type, as demonstrated in Fig. 9.55.

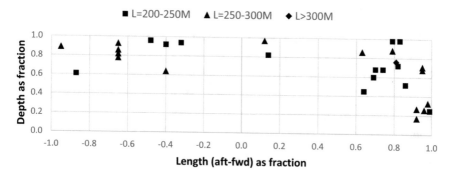

Fig. 9.51 Renewal locations of side frames i.w.o. side shell along length (P)

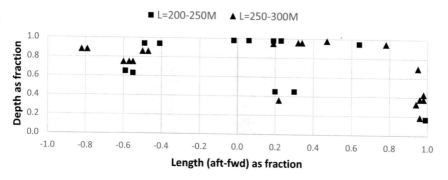

Fig. 9.52 Renewal locations of side frames i.w.o. side shell along length (S)

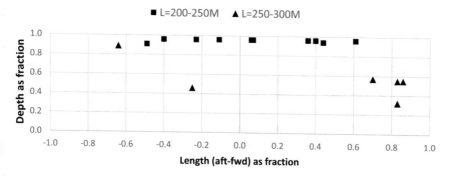

Fig. 9.53 Renewal locations of side frames i.w.o. longitudinal bulkhead along length (P)

9.5   General Conclusions

The analyses of renewal locations of selected steel structural members reveal some basic facts. Renewal locations of deck plates, deck longitudinals and deck transverses

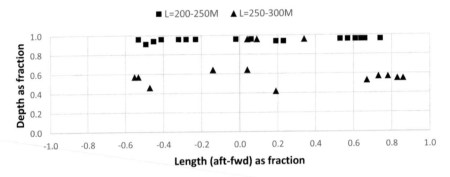

Fig. 9.54 Renewal locations of side frames i.w.o. longitudinal bulkhead along length (S)

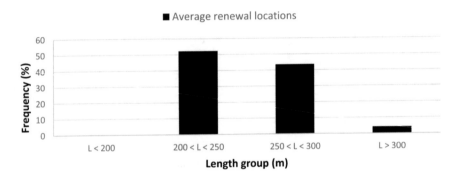

Fig. 9.55 Number of renewal locations versus length group

(Figs. 9.1, 9.8 and 9.18, respectively) are located mainly at the extreme transverse end of the main deck (port and starboard). Similarly, renewal locations of side shell plates, side shell longitudinals and side frames (vertical transverse webs on side shell) (Figs. 9.4, 9.11 and 9.21, respectively) are located mainly at the top areas of side shell plates (port and starboard). Likewise, renewal locations of longitudinal bulkhead longitudinals and side frames (vertical transverse webs on longitudinal bulkhead) (Figs. 9.15 and 9.24, respectively) are located mainly at the top areas of longitudinal bulkheads (port and starboard). The analyses also suggest that the age and length do not directly impact these renewal locations.

9.5.1   Renewal Locations by Dimensions

Figures 9.1, 9.2, 9.3, 9.4, 9.5, 9.6, 9.7, 9.8, 9.9, 9.10, 9.11, 9.12, 9.13, 9.14, 9.15, 9.16, 9.17, 9.18, 9.19, 9.20, 9.21, 9.22, 9.23, 9.24, 9.25, 9.26 and 9.27 demonstrate the renewal locations of various structural members. The results of the analyses of renewal locations show that the renewals of deck plates, deck longitudinals, deck transverse, side shell plates, side shell longitudinals, longitudinal bulkhead longitudinals, side frames (attached to side shell plate and longitudinal bulkhead) are concentrated in the vicinity of the intersection of the main deck and side shell. In other words, deck plates with attached members at the extreme end of the main deck, side shell plates with connected members at the top and longitudinal bulkhead with connected members are the areas where maximum renewals were carried out irrespective of age and length. As such, these are the areas prone to structural steel replacement.

The ship's main structural members are deck plates, side shell plates (port and starboard), and bottom shell plates. In real life, these plates, alternately, experience tension and compression due to the ship's movement—hogging and sagging together with rolling, pitching, heaving, and so on. Therefore, it is very much expected and supports the theory of longitudinal stresses in an inclined condition. The analysis of the longitudinal stresses under hogging and sagging in an inclined condition suggests that the maximum and minimum stresses of an element in a section are associated with the distance of the element from the axis, and that means that these stresses will occur at the corners of the transverse section of a ship. In other words, these four corners will suffer fatigue failure, mainly, top corners (port and starboard), coupled with corrosion effect due to the presence of vapour/gas at the top part of the compartment. It is to be noted that geometrically, the vertical position of the longitudinal axis through the centre of gravity is less than half of the depth of a ship and tanks are never loaded 100% to its volume (typically 98% maximum), allowing to accommodate gases and vapours which accelerate the corrosion process.

Under the inclined condition, the following relationships are developed and used to calculate the bending stress [4, 5].

$$M_y = M\cos\theta \tag{9.1}$$

$$M_x = M\sin\theta \tag{9.2}$$

$$f = \frac{M\cos\theta}{I_{NA}/y} + \frac{M\sin\theta}{I_{CL}/x} \tag{9.3}$$

where

M: the bending moment in the vertical plane

θ: the angle of inclination

M_y: the component of M in the vertical plane under the upright condition

M_x: the component of M in neutral axis under the upright condition

f: the bending stress under the inclined condition

I_{NA}: the moment of inertia about the horizontal axis under the upright condition

I_{CL}: the moment of inertia about centreline under the upright condition

x, y: the coordinates of the point (under the upright condition) of which the stress to be calculated.

Equation 9.3 also suggests that higher x and y values yield a higher stress level. According to the ship's geometry, the intersection of the main deck plate and side shell plate (port and starboard) possess the highest value for x and y from the longitudinal neutral axis. Also, component $\cos \theta$ yields the highest value at zero inclination (at upright condition), and component $\sin \theta$ yields the highest value at 90° inclination. In real life, the maximum and minimum stresses occur at an angle of about 35° [4, 5].

Figure 9.56 is the hypothetical cross-section view of combined deck plates, side shell plates, deck longitudinals, side shell longitudinals and longitudinal bulkhead longitudinals. It confirms the same fact and demonstrates that the maximum number of repair locations of deck plates, side shell plates, deck longitudinals, side shell longitudinals, and longitudinal bulkhead longitudinals were taken place in the way of the deck (extreme port and starboard) and side shell (top corner area close to deck plate).

Longitudinally, renewal locations are distributed all over the length with a different pattern for different structural members, and they are more in the forward of midship than the aft. Figure 9.57 demonstrates the same facts. It is constructed by combining Figs. 9.1, 9.4, 9.8, 9.11 and 9.15. The different concentration of renewal locations is observed at the forward (about 25% of the length from forward perpendicular). It is, possibly, due to the damage caused by panting and pounding effects. Also remarkable is very little concentration at the aft (about 20% of the length from after

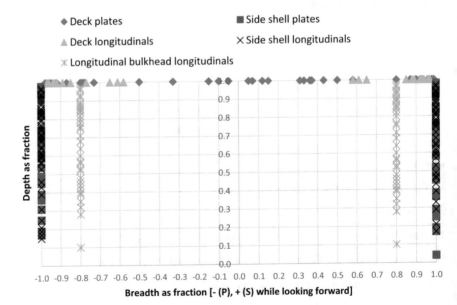

Fig. 9.56 A typical midship section with steel renewal locations of selected members

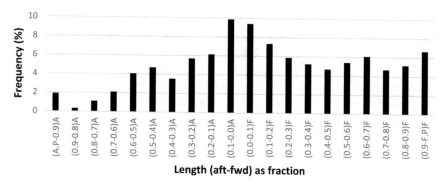

Fig. 9.57 Distribution of renewal locations of structural members under study along length

perpendicular). It is probably, due to the profile of the hull bottom around the propeller area.

Around this area, waves move/roll over the surface with less impact, unlike a massive impact at the bow area. Moreover, the aft area, including the engine room bottom, is stronger than the bow area. This is due to the presence of the main propulsion engine and other heavy machinery. Classification Society rules also require more massive construction around the engine room and stern area.

9.5.2 Renewal Locations by Age Groups

The analyses of renewal locations by the age of a ship show that the age of a ship does not have any direct relationship with the renewal locations of structural members. However, most repair locations lie at the extreme transverse end of the main deck (port and starboard) for deck plates and their attached members and at the top corner of the side shell (port and starboard) for the side shell plates and their attached members. In addition, age groups of 10–15 years and 15–20 years suffer from maximum renewal locations (Fig. 9.41) irrespective of their length, deadweight, and type.

9.5.3 Renewal Locations by Length Groups

The analysis of renewal locations by ships' length shows that the length of a ship is not directly ed to the renewal locations of structural members. However, most repair positions lie at the extreme transverse end of the main deck (port and starboard) for deck plates and their attached members and at the top corner of the side shell (port and starboard) for side shell plates and their attached members. Length groups of 200–250 m and 250–300 m suffer from the maximum renewal locations (Fig. 9.55) irrespective of age and type.

Finally, the locations of structural steel renewal are influenced neither by age nor length of a ship. Still, renewal locations are concentrated at the intersection of the main deck and side shell with their attached members (port and starboard) due to ship's movement mechanism. During the internal inspection, these areas need more careful attention to avoid any disastrous consequences without an alarm (Fig. 9.56).

Mentioned structural members and areas are in the way of wing tank—port and starboard side. Therefore, during tank inspection, by the shipowner or by classification society's surveyor, special attention must be paid to deck plate with deck longitudinal and deck transverse, side shell plate with shell longitudinal and side frame (vertical transverse web on side shell), longitudinal bulkhead longitudinal and side frames (vertical transverse web on longitudinal bulkhead). Those are mainly wasted in scantlings.

References

1. Dev, A.K., Saha, M.: Modeling and analysis of ship repairing labor. J. Ship Prod. Des. **32**(4), 258–271 (2016)
2. Dev, A.K., Saha, M.: Analysis of structural steel renewal locations in ship repairing. J. Ship Prod. Des. **37**(1), 1–36 (2021)
3. ISSC: Report of Committee V.2: Inspection and Monitoring, Chaired by Professor George J. Bruce, Fifteenth International Ship and Offshore Structure Congress, 11–15th August 2003, San Diego, USA (2003)
4. Muckle, W.: Strength of Ships' Structures. Edward Arnold (Publishers) Ltd., London (1967)
5. Robb, A.M.: Theory of Naval Architecture. Charles Griffin & Company Ltd., London (1952)
6. Seref, A., Mirza, A.: Bulk carriers repair analysis. Ann. J. Soc. Naval Arch. Mar. Eng. **30**, 51–61 (2008). ISBN 978-981-4222-97-6
7. Smith, C.B.: Extreme waves and ship design. In: 10th International Symposium on Practical Design of Ships and Other Floating Structures, September 30–October 5, Houston, TX (2007)

Chapter 10
Data Collection and Management

Management information is an essential component of any business's successful operation and is as complex as ship repairing, including the international supply chain, it becomes critical. Management can be defined in many ways, but a helpful definition includes "co-coordinating groups or individuals towards a common goal" and "organising, coordinating and controlling resources to achieve an objective". Both statements fit closely with the need for good ship repairing project management.

Information can be defined as "processed data". Raw data, collected from the plans, schedules and past operations records, is not necessarily beneficial for management. The data can be sparse or excessive and is often unstructured. Processing the data puts it in a helpful format by identifying anomalies that require some investigation or detecting trends that may be good or bad.

Information can reduce uncertainty, giving management confidence that the actions they plan to take are the best ones. Information can also be false or inaccurate, perhaps if a third-party stakeholder provides it in a project. It can be new or incremental, adding to the knowledge of a management team. It can confirm or sometimes correct past information.

A system can be defined as "a set of interrelated components". The components are interrelated for a purpose, which for a company is to achieve the objectives of the management. The company's management has overall corporate goals, usually starting with an overall goal of being sufficiently profitable and still being in business in five years' time. To achieve these, the immediate aim is to complete a successful ship repairing project.

All the necessary control is managed through information. The information must be of good quality so that the management has a factual basis for making decisions about sub-system changes. The information's timing is essential because it must be available in sufficient time for effective action to be taken. The information must have relevance, so it deals with meaningful issues, such as deviations and does not offer unimportant trivia. (see Hills, et al. [1])

A. K. Dev et al., *Ship Repairing*, Springer Series on Naval Architecture, Marine Engineering, Shipbuilding and Shipping 12, https://doi.org/10.1007/978-981-16-9468-4_10

The presentation of the information is essential because it must be readily understandable to the recipient. Appropriate detail must ensure that the recipient has enough, but not too much, which can obscure the issue. There must also be a good volume of information, enough to inform decisions and no more. Extracting meaning from a lot of figures is very difficult if there is limited time to study them. Having information about activities completed on time and correctly can distract a user from the few variations essential to correct. Graphical presentations can be helpful, but users may need specific training in their correct interpretation.

Once the data has been collected about whatever situation is to be studied and processed into useful information, it must be communicated to the users. Thus, there is a necessary sequence of actions to carry a piece of information from its source to a manager (for example) who needs to act.

First, an observation must be made. This can be to complete a production process, pass a piece of work by an inspection, note an item has arrived from a supplier, and any other event. In general, these events will have been planned. On the other hand, an observation can be of the failure to complete something, so the event does not occur.

The observation then must be understood. It is always possible that a mistake has been made in the event observation. This results in an item of data, correct or otherwise, which needs to be communicated to a supervisor, manager, or staff member charged with recording events. To do so, it may need to be encoded. This could be by word of mouth, in writing, electronically or visually.

The encoded message is then transmitted to the recipient. In the transmission, the message can be distorted by a verbal message obscured by noise. Likewise, radio communication can be distorted, and other problems may arise.

The message then reaches the receiver, possibly the wrong data or distorted or both. Then, for example, it must be decoded, transcribed into an IT system, or otherwise organised to act.

At any stage of the communication process, therefore, it is potentially possible that errors will occur. Errors that arise should be investigated, and some action is taken to modify the communication system to avoid future similar problems. In addition, all data which may be critical should be checked for potential errors.

Data collection as described is fundamental to effective operations. Moreover, once available, the data can be turned into useful information. This is needed for several reasons.

First, to manage the ongoing operations in the shipyard. This requires direct knowledge of events that occur, mainly if any delay or other problem occurs. Here, the need is also for accurate information to assess any situation and develop a recovery plan.

Second, for the planning of future operations for current contracts. Again, this requires data taken from past agreements. Although there are often significant variations in the work content and tasks from contract to contract, what happened in the past can provide a valuable basis for planning.

Third, the above generate the requirements for analysis of past contracts to improve estimating and planning for future work. Here the data can be reviewed and analysed to offer suitable information for the future intentions.

At the point in time when an event occurs, data about that event can be perfect. All facts about the event are clear. For example, if a worker has completed a task, the worker's identity, time is taken, the status of the item worked on, location and any other details are known. If that data can be captured at that point in time, then there is a complete data set. However, if the data collection occurs later, for example, at the end of a working day or when an inspection takes place, the data may not be complete or accurate. Thus, at its simplest, the memory of events may be mistaken.

There is always a danger that data will degrade over time. So ideally, the data should be recorded immediately on completion of a task. This can be difficult, for example, in a dirty or cramped space, where the workers must clean themselves and their equipment, or when there is another urgent task to start. This may lead to reliance on another, for example, a supervisor, to make the necessary record from a verbal report. This requires the supervisor to be available immediately. The information needed to be collected can be extensive, including the worker hours used and who did the work by trade, especially if there are hazardous conditions. There is also an increasing need for regulatory information, for example, on environmental issues, including waste management.

Suppose an inspection is required from a supervisor, shipowner representative or external authority, then the completion must be reported, followed by a delay until the inspection can be carried out. In the meantime, the workers cannot be left to wait but need to be moved to a new task.

Ideally, the data will be recorded automatically, as far as this is possible.

The software systems which are used to operate any business, including enterprise resource planning (ERP) and management information systems (MIS) in general, rely on production data collected primarily from the shop floor. For example, the shop floor for a ship repairing operation includes onboard ships under repair, other work sites at quays, docks and around the often-extensive area. Naturally, this increases the collection problems. In most cases, this data is collected manually and entered the system by a human operator. (See Brewster, et al. [2])

The three common problems with manually collected data are untimeliness, inaccuracy, and bias due to the communication and other problems outlined earlier. Considering that this raw data forms the basis for all subsequent production reports—and that important decisions are made based on those reports—any problems with the initial data collection can start a ripple effect that negatively impacts a business.

In a typical scenario, manually collected production data is entered into the system only at predetermined times, such as at the end of a shift or job. The information then is made available in the form of reports and used for analysis. For long-term study, viewing the data after the fact is usually sufficient and valuable for future reference. However, ship repairing is a dynamic environment, and knowing what happened only sometime after the fact can seriously detriment productivity.

Manually entered data also can be incomplete. This is because the actual data entry task often falls to someone with many other responsibilities. In addition, data entry is a tedious task that often is put off for as long as possible. It is not uncommon for raw data to sit around for hours, if not days, before entering the system. As a result, reports don't show the latest data. Further, given the unwillingness to complete data entry, there can be neglect if some data never seems to be used afterwards. As a result, it is common to find incomplete records when the past data for a company is examined.

Using dedicated staff for data entry is used as a solution. However, the process is still dependent on the generator or current owner of the data (say, a supervisor, passing the data to the entry staff as soon as possible). In addition, transcription errors have been referred to, and this method offers an opportunity for such mistakes to occur.

Often the data must be written down first and later entered the system, sometimes by a different person than the one who recorded it in the first place. Typographical and transcription errors are common. Once these errors become part of the data set, they become difficult to detect and eradicate, making all the resulting reports suspect.

In addition, a human operator influences what information is entered and when.

Collecting production data automatically, as it happens, can help eliminate these problems in some industries, but the potential in ship repairing is limited. Until recently, commercially available data collection software tended to be vendor-specific, designed to collect data from proprietary equipment. As a result, for most ship repairing tasks, data collection remains a problem.

Recording production data is first about completing work tasks or, more importantly, recording delays in the tasks. The first pre-requisite is to have workers aware of the importance and supervisors to whom they report redeploy workers when any task is finished or stopped. This will ensure that data is identified as quickly as possible so it can be recorded. (See Lindahl [3])

The recording should be a management information system, where all the tasks have been identified and the necessary resources and timings included in the task definition. This can be done by word of mouth to data entry or, ideally, directly by the supervisor. The task can be completed, and then the next task for the workers is allocated. Or if it is stopped, this is flagged, so the task is highlighted in the system as requiring attention.

It may be possible to wait until the next contract progress meeting to report and identify the next step to bring a task back. However, if it is urgent, a conversation will be held immediately and probably a meeting at the worksite to make a full assessment.

The information system will act as a permanent record of all events, and ideally, the task will have a red flag so that scanning the task schedule for the contract will alert management to the problem. Ideally, also the local supervisor can update the records in the system. The data entry should be as simple as possible, and since the task is fully defined in the system, all the supervisor needs to do is amend the task status.

Some shipyards employ progress chasers who move around a ship, or ships if they are small, to note the status of each live task. A combination of people on the ship under repair and a recording system, so status is available to more senior management, should ensure work progresses and all workers are fully used.

Within a workplace, employees need to be up-to-date with the risks, hazards, safety measures, and emergency procedures that are in place. Most employees should be offered regular safety training and further training when new equipment is purchased or working practices change.

A record of training undertaken should be kept within the overall information technology system which is adopted by a shipyard. The data will be maintained alongside records of work undertaken, mainly if any is carried out in hazardous locations. Data will include the employee personal details, role, skills, work histories, disciplinary action if any has been taken and any future requirements or plans. In addition, accurate training records will keep track of when employees need to be retrained.

Specific workplace injuries, accidents, or near-misses need to be recorded, a statutory requirement in most countries. Data required will be all the details about an incident, and this data can be analysed later to create reports that help identify patterns and areas that need improvement. The data will be transferred and stored securely, and access to the details can be limited to relevant staff members.

Staff should be encouraged to report all safety issues even where no actual incident occurs. Any liquid spillages, loose cables, trip hazards, debris build-ups and other similar occurrences can be dealt with to create a safer work environment and reduce waste or lost time. In addition, good data will allow the management to identify and manage any weaknesses in their operating methods.

It is essential to ensure that the records kept are always accurate. Avoiding a blame culture will encourage correct reporting, and then the data collected can be stored securely and efficiently accessed when needed.

Data collection can be automated to some extent, and this is an excellent way to reduce the possibility of incorrect information entering the system. For example, the presence of workers and sub-contractors in the shipyard can be recorded using individual smart cards detected on entry. Linking the cards to payments and safety information is an excellent way to ensure they are looked after and managed correctly.

The cards can then record the presence of a worker at a specific worksite, which goes a long way to record the work they are engaged in. If the worksite is aboard a ship, then the safety aspect becomes essential and as this can give a secure and accurate check on who is aboard in case an evacuation is necessary. This is particularly important for fire safety.

A specific worker in a known location onboard a ship will almost always identify the work being carried out. The task descriptions for each work package can be encoded and then tagged to all materials required. The supervisor assigning tasks can record the workers assigned, giving a link to intended and actual presence. Work hour recording is then more or less automated, with the supervisor checking any anomalies.

Materials and equipment arriving in the shipyard can be tagged with their details before any storage or use. This also makes the collection of actual costs easier once the items are assigned to a work package.

Good data collection depends on management in a shipyard seeking to collaborate with the workforce to mutual benefit. If workers are regarded as cheap and expendable, then their cooperation is unlikely. Workers who are not well treated will always find ways to avoid a management system that they do not trust or where recording any mistakes or problems is expected to lead to problems. Some ship repairing companies are known for reacting to situations by firing people to encourage others. This attitude rarely succeeds for any length of time as workers will leave if they can or withdraw cooperation if they cannot.

Data is only as good as the accuracy of its collection and the careful use for analysis. If this is neglected, then a shipyard will not be as efficient as possible, and there may be serious problems with severe consequences.

References

1. Hills, W., Snaith, G.R., Bruce, G., Braiden, P.: Computer-Aided Engineering in Shipbuilding and Repair, Final Report, Department of Trade and Industry, Contract Report (1992)
2. Brewster, A., Bruce, G., Evans, M., Hills, B.: Development of cost-effective computer management information systems for small shipyards. J. Ship Prod. 13(2), 125–137 (1997). The Society of Naval Architects & Marine Engineers, Jersey City, NJ, USA
3. Lindahl, I.: The computerisation of the information flow in ship repair. In: 8th International Conference on Computer Applications in Shipbuilding, Sept 5–9, 1994 in Bremen, Germany, vol. 2, Berry Rasmussen Reklam AB, Sweden (1994). ISBN: 91–630–2762–3.

Chapter 11
Management Information Systems

Planning and control of ship repairing are often inhibited by the inherently volatile nature of the business. There are inevitable, short-term changes in work scope as contracts are estimated, secured, and then progressed. Management systems must differ from most industries, including ship construction, in that they must be designed to cope with changes. For most businesses, change and a wide variety of items to manage are exceptions.

Ship repairing has been variously described as "complex", "dynamic", "fast-moving", and "chaotic". The business of ship repairing is undoubtedly inherently challenging to plan and then manage. There are some very different categories of ship repairing and conversion to prepare a ship for some new service. A ship repairing contract can vary from a few hours to days, weeks or even months. The main types of work can be summarised, although there is a continuum of work types that do not fall precisely into fixed categories. The work types identified are:

Voyage repairs, where minor repairs are carried out with a ship in service. This is often away from a shipyard site, with a mobile work squad operating in a port environment.

- Routine drydocking, where the ship is docked for hull coating renewal and any other required underwater work when the opportunity is usually taken to make additional repairs. Such drydocking will coincide with regulatory checks on a ship.
- Drydocking for a special survey, at five-year intervals, when steelwork renewal may be needed, especially for some types which carry aggressive cargoes. The scope of the inspection and potential repair will increase as a ship ages.

Damage repairs, where extensive work, particularly to the ship's structure, may be required. This can occur at any time and is usually unpredictable, especially if a ship cannot be temporarily repaired and then taken to a preferred shipyard.

Conversion is where a ship is refitted for a different use or upgraded to give a life extension. The work is often extensive, including accommodation, engine replacement, lengthening and other systems upgrades.

The planning and management needs vary according to the category of work and ship type. For example, significant conversions and refits are considered closer to new construction in organisation and management terms, but what management systems are available to support routine and emergency repairs? This is a particular problem for the smaller companies, which have limited management resources.

Repair work creates more problems than ship construction. First, the forward workload is often not specific at the time of commencing the contract work. Second, although a specification for the repair work will be provided by the shipowner or superintendent of the ship, this is often only a starting point for the work. Unexpected items can be found which have a higher priority than the work specified. The shipowner may reduce the budget for the repair work. The work may also be curtailed because the shipowner receives a new, urgent charter.

A conversion usually includes some initial uncertainty about the existing ship condition, which is often treated as a separate repair contract. The ship is then in an acceptable condition for future operation, and after that, the actual conversion work can be primarily planned and managed in the same way as a new building.

Second, after inspection, more work may be required as the contract progresses than was initially envisaged. This may sometimes lead to the cancellation of other, less urgent jobs to allow the total cost of the repairs to remain within a budget total.

A third complication is that unlike ship construction, where a cost estimate for the contract is based on productivity measures for past work, most reasonably routine, the workload is variable. Only limited measures for productivity are available in ship repair.

If a shipbuilding contract requires additional man-hours to finish, usually the shipyard bears a loss. However, in ship repairing, the requirement for additional man-hours may increase turnover and profit. (See Stewart [1])

Most other industry sectors selling services and repair facilities have non-unique products or can allow completion dates to slip. As an example, road vehicle maintenance is generally carried out at a central depot, and if additional or more challenging repair requirements are identified, another vehicle can be substituted until the work is complete. Other industries do not operate in an international trade, where even if another ship might be available as a substitute, it is likely to be too far away to assist.

The volatility of ship repairing makes demands on the management information systems (MIS). Management is often conducted relatively informally, so decisions are taken locally to expedite work processes with formal paperwork catching up later. To keep control as the contract progresses, it is essential to collect accurate progress information and transmit this rapidly to the contract management team. Almost real-time data collection is desirable to be able to manage rapid changes in work scope.

Any change in work or schedule in ship construction is avoided, if possible, whereas in ship repairing, change is inevitable. This can leave conventional planning software struggling to keep up with actual progress on the ship.

Accounting software provides an accurate cost outcome (subject to the input information being correct) but is essentially designed for historical cost recording. It is not usually intended to give the flexibility in updating, particularly variations in work scope and task cancellations that ship repair requires.

There is a need for an integrated system to manage the specific problems found in a ship repair contract. There is a further need to look across all the current contracts in the ship repairing yard and consider any enquiries that may be converted in the period covered by the shipyard planning horizon. The rapidly changing contract situation, and the possibility of a damaged ship requiring immediate attention, also contribute to the volatility of the management situation.

The specific needs of the ship repairing industry must direct the development of any effective management information system. These needs can be summarised:

Marketing must provide the shipyard with a stream of "good" enquiries. A good enquiry is one where there is a good chance of a contract with a reasonable profit. To achieve this, there must be up to date and comprehensive management of the client base. The clients are likely to be international, even for a small shipyard. Their need is to track the companies, agents, management companies, shipowners, technical staff, superintendents, and ships that may need repair. The data on all these must be accessed and updated remotely, so a shipyard manages rapidly changing circumstances. (See Evans [2])

Tracking the ships is vital as these will often change ownership and name. Tracking people is also important because when they relocate, there may be additional opportunities for a shipyard.

Many data sources can be accessed, including vessel tracking sites, port operator information on ships in port and due, records of ships data which will provide intelligence on potential dates for regulatory inspections and the shipyard's records of past enquiries and contracts. All of these and personal contacts build a picture of potential future activity that the shipyard may access.

Frequent contacts with shipowners and operators can be made and recorded to gain information and ensure that the shipyard's capabilities are known. Taking a longer-term view, the shipyard may attend international exhibitions and conferences to promote further the services offered and the experiences on which these are based. Direct visits to potential customers can also be made and using the data available additional potential customers can be identified and visited while company representatives are in their vicinity.

Management of enquiries once received is the next stage of the process. The typical repairing contract can be relatively small, and the success rate on enquiries is usually no better than one in five. This may increase to one in ten successes in a poor market, which presupposes that the selection of initial inquiries is well managed. A significant number of live inquiries must be managed at any time, linked to the client database, and the status of any single enquiry must be updated frequently.

In contrast, some ship repairing companies have long term arrangements with one or a small number of companies, an example being the UK repairers and ferry owners operating between the UK and Europe. Others specifically look for larger-scale contracts, including the modernisation of ships and even conversions. Changes in international regulations are another potential source of a stream of contracts, for example, ballast water management and other "clean" technology updates.

In all cases, the management requires a database of current agreed contracts, which may be only a few weeks into the future, although some longer-term commitments may exist. Agreed future contracts may be lost if a ship is diverted to an alternative route, so it will no longer be near the shipyard when the repairing work is required. The database then collects all possible contracts based on the current enquiries and on the marketing effort. The data required includes any intelligence on the ships and their owners, their trading patterns, and the shipping market's state. Past contracts with the same shipowners and other experiences, perhaps where enquiries have been received but never been converted to contracts, are also important.

From an analysis of all the available data, some estimate of the probability of success with enquiries must be made, which will inform the management on which inquiries should be pursued vigorously and abandoned. Further decisions can also be made on the prices to ask, and perhaps whether to offer a small discount as an incentive. Enquiries can be ranked based on the information generated.

The enquiry database will provide as comprehensive a picture of a company's likely future as possible. Management judgment is still required, and there is often potential for a contract to appear suddenly if a ship is damaged or has a severe breakdown near the shipyard location. So even weather forecasts can be part of the overall picture. Once the database has been established, it must be maintained, but a well-designed system will allow any data to be entered quickly and easily. It can then be retrieved and combined with other information to enable a helpful picture of the possible futures. The information must be accessible remotely, for example, to support shipyard staff travelling to visit potential customers.

Man-hour estimates are ideally based on some, however, limited, analysis of past performance so that they can be "realistic". A realistic enquiry is one where if the contract is won, it will be profitable. How the estimating process is managed can vary between different shipyards. It can also depend to some extent on what the owner requests as part of the enquiry response. (See Bruce [3])

Most shipyards publish a tariff of costs for specific tasks, for example, pipe replacement, generally based on the location and dimensions of the pipe required. However, realistically, these do not fully reflect the actual cost of the task to the shipyard, ignoring some aspects of the condition of the pipe and the space it is in, which can affect access and difficulty of removal.

Many shipowners ask for rates for tasks, such as the cost of replacing a kilo of steelwork. Again, although these are based on the total quantity of steel and the location, they can only offer a rough guide to the actual work costs. As a result, some shipyards use an internal, unpublished set of rates for tasks as the basis of their estimates.

It may be necessary to maintain several alternatives to allow a shipyard to respond quickly to an initial enquiry. The shipyard should also carry a database of past contracts with the prices charged and the actual costs of the work as recorded at the time. The prices and costs are very likely to be different, but the estimating process must build. Some owners dislike certain charges, often those for fire, safety, service provision and drydock rent. These are charged daily and daily, and to win a contract, a shipyard may have offered these at a reduced charge. Such prejudices can form part of the data on particular owners.

Ideally, the shipyard will record the real costs of work done, including the actual work hours spent on service tasks such as access equipment or scaffolding which may be spread across several work tasks for billing purposes. However, the shipyard costs may not correspond to the charges to the owner, which will generally be based on the specification items in the original enquiry. These may be modified to reflect shipowner prejudices and for commercial gain.

So overall, a database is needed that can retain several versions of contract costs, which can provide a valuable basis for future estimating.

Materials and supply estimates are built up over time using readily accessible past data, supplier estimates of cost, sub-contractor rates and standard tariffs. Rapid and easy access to past estimates and contract data is valuable to the process. However, market prices can be very volatile. For example, steel prices have fluctuated in recent decades by over one hundred per cent, and expenses of other materials or equipment also vary dependent on the overall marine market. Sub-contractors will have to estimate their costs to the shipyard and have similar difficulties.

The database has to include intelligence on suppliers and contractors like that on ships and their owners. Frequent contracts will require regular suppliers, which is helpful both to support the shipyard and give the suppliers some indication of the potential future shipyard requirements.

The contract negotiations can be protracted, with changes in specification and work scope. However, easy tracking of the changes and status increase the chances of success. This can be managed through an effective estimating database, recording all changes in the work scope, and re-negotiating prices.

Once an estimate has been converted to a contract, production planning must convert that estimate into a work schedule, which is then maintained in an up-to-date form as the work scope changes. The schedule is often variable, as new work emerges and tasks may be cancelled or reduced in scope. The planning will also convert the work hour estimates into a form that relates to the management of the work. Costs from individual tasks from the owner specification can be consolidated to provide budgets for different trades, mainly services. The final costs will then be reallocated to the shipowner tasks when the invoice for the contract is presented.

The work schedule is also the basis on which to record actual man-hours, and other costs and also to track progress. This allows the status of a contract to be assessed quickly, along with its potential profitability, at any time during the duration of the work. It also allows easy comparison with the estimate to monitor changes. Control of the operations is essential to ensure that the contract outcome is as predictable as possible. Any potential problems, mainly affecting delivery, are flagged early so action can be taken.

Purchasing is frequently part of an accounting software package because it is a high cost in ship repair. It is, however, essential that the status of any orders placed for equipment spares or materials can be accessed quickly and accurately so that management can plan the work tasks that require them. It is also very desirable to have a fast track in the information system. An order can be placed immediately if a new material or equipment need is identified during the contract. Occasionally, a new supplier may be required, and some accounting packages can be challenging to use flexibly. For example, entering details for a new supplier can be slow, and big data may be necessary to place an order.

Often a telephone order needs to be placed, in which case the information system must be able to accept minimum data to record the order. Then this is flagged on the system to ensure that the verbal order is followed up. The ability for a system to be flexible in operation is essential to allow for the fast-moving nature of a ship repairing contract.

Once an order is placed, the management must follow progress, and where necessary to expedite delivery. The information system requires features to make this possible, including records of conversations and other contacts with the supplier.

Once a delivery has been received, the system should record that it has been checked for quality, quantity, and location in the shipyard. On occasion, an urgent item may be delivered outside regular working hours. In such a case, a feature is required to record the item's arrival, even if there is no complete set of details about the item and the supplier. For example, a security guard may accept the item when no one responsible for the work task linked to the item is available. Then as much information is recorded, the item is placed in a secure location to wait for inspection.

An item accepted into the shipyard and not required immediately must have a storage location on the information system to be made available when needed for production.

As each task is started and then completed, progress, materials, equipment, and use of resources are recorded on the system to track progress. Work hours can only be recorded against a task that has been opened for work. Once a task is registered as complete, no additional hours can be recorded. If this is not managed carefully, hours can be misallocated, whether accidentally or deliberately, potentially costing the shipyard.

With the hours and costs associated with each task, the completed work schedule should lead to an accurate invoice. The outcome can be compared with the estimate, extras highlighted, and client agreement can be managed more effectively. Maintaining up to date and readily accessible contract information benefits both shipyard and shipowner.

The data from the contract management process includes information required for other functions in the shipyard, for example, management of sub-contractors. Accurate man-hour recording against tasks provides the necessary information to support the payroll, whether internal or sub-contracted. This is also accessible for a future estimate, as is the estimated and then final costs of bought-in items and services.

An effective and integrated system can be developed by building a series of standard processes around a central database. With all data residing in one place, accessible and flexible reporting is possible. Current and historical data can be aggregated, allowing a corporate 'memory' to be built up. The data can be consistent and thorough, preventing errors and trapping all costs.

Having specified how an "ideal" management information system would support a repairing shipyard, however, there are multiple obstacles to effectively managing ship repairing using such a system. As has been mentioned, management information is only as current and as helpful as the data entered. Existing software may permit swift, efficient reporting but often has cumbersome data entry procedures, mainly where changes occur. This is often the case with accounting software, where the aim is to obtain accurate costs rather than facilitate the contract completion. The quality of the data is generally inversely related to how painful it is to enter. Inaccurate information is positively dangerous to the company.

Often those charged with entering the data perceive little value for themselves from a system. They view such effort as interfering with their 'real' work, or worse, an irrelevance. In contrast, those who extract information from a system can significantly benefit and place ever greater reliance on it. This gap in user 'buy-in' hinders the inevitable cultural changes required if a system is to be a success. Ease of data entry is critical to a successful management system to reduce the potential for errors in the data entered. Poorly designed user interfaces can result in more significant error rates. The system must also be able to manage large volumes of data from multiple users.

Computer systems are very good at accelerating repetitive and predictable processes but often implicitly acknowledge that events that fall outside routine will be managed manually. This is not sufficient in ship repairing, where even a single error or missed cost can turn a contract's profitability. Every time a situation is beyond the system capabilities, the system is devalued. In looking at data collection, the point has been made those information systems often do not capture non-standard items. Many systems in a range of industries have unofficial workarounds where non-routine data is found. For example, unofficial spreadsheets usually exist alongside the official company systems. An adequate system must treat the non-routine as expected for a successful ship repairing operation.

Management information systems use some fixed model of the business they manage. It results from a systematic analysis of the company to be managed, usually conducted by a software developer. Once complete, the model is set, and as the information system is developed for it, any changes become increasingly difficult. Often the need for a change in the company model is only identified after the software is partly written and testing is in progress. As a result, it is easy for analysts to miss aspects of the company operation, especially if informal and unrelated systems are used. It is also tricky for company staff to identify needs for a system that is not yet developed when they are asked.

The company's developed model can either be generic for an industry or can be expensively designed for a single shipyard. In an industry as flexible and changeable as ship repair, the former leads to compromise and the latter too enormous development costs. The initial cost of a software system is rarely the final cost to the user. Redesigns of system elements and updates to fix problems or missing needs can last for years and be expensive. Realistically, most software can be broken into small segments that apply to many if not all systems. How they are assembled for a specific company is where the differences in final results will exist.

The effort should be spent on a comprehensive understanding of company operations, with all staff involved in initial consultations and reviewing what the system will look like to the users. Many of the users will not be previous users of IT systems.

Traditionally management systems have a static view of the world. The data structures are created according to a snapshot of the business at a single moment in time. This can be misleading and inappropriate. For instance, marketing data is often an extended address book with a strict hierarchical structure of companies, employees, and ships. However, a shipyard operates in a dynamic community where relationships between companies, people, and ships continuously change.

There has been a traditional separation within computing between database data, electronic files such as spreadsheets and word-processed documents, and, importantly, e-mail. As a result, the database itself rarely holds all the information pertinent to a contract. And this is without considering paper records, notes of telephone conversations, e-mails, and other less official contacts.

Management information systems may be seen as the first step to an electronic marketplace. Ship repairing relies upon practical and remarkably sophisticated supply chains, but electronic data interchange within ship repairing has yet to deliver. It is hard to beat the efficiency of the traditional phone call to a favoured supplier, avoiding proprietary databases, "harmonised" product catalogues and security concerns. This aspect has to be captured in some form for a system to be fully effective.

Most companies in the ship repairing business have developed an in-house estimating process. There is no equivalent of the parametric estimating available for new construction. If repairing costs are plotted against ship parameters, some correlation can be found, but the variability between contracts is so large that the information is of no value for commercial estimating. It can be shown that the typical growth of a contract from estimate to final invoice is around 20% in many shipyards, but again any individual contract can vary considerably more. The variation can be lower as well as much higher, so again the information does not help forecast an individual contract.

The estimating process is generally based on a spreadsheet developed by the estimating personnel and gradually adapted over time for smaller companies. Such spreadsheets are effective in carrying out the primary estimating task and are adaptable to new requirements. On the other hand, this approach can lack internal consistency as the spreadsheet changes over time. For example, individual estimators may make personal changes to suit their current projects, which others may not capture. In addition, when estimators move companies, much of the personal knowledge may go with them. As has been stated, the basic process of developing an estimate is relatively simple, and a spreadsheet can manage this. However, the capture of expert knowledge and the ability to readily use past information within a group of estimators is often missing.

The spreadsheet will list the tasks to be carried out. For the shipyard, some tasks will correspond with the shipowner requirement directly, for example, inspection and repair of a piece of deck equipment. In other cases, such as replacing damaged plating, the task may include scaffolding or other access equipment and several trades. Some support trades such as scaffolding may be used for several tasks in the same part of the ship. To calculate the estimated costs of these tasks, the estimator needs to break the costs down, usually by trade. The estimate can be created using the shipowner's job list as a basis, but this creates problems in accurately assigning man-hours for some work, especially the support tasks.

The estimate created to win a contract will form the basis of the job list for the production workforce. If it is based on the owner jobs, then the correct assignment of man-hours also becomes a problem. Ideally, the estimate will manage both the shipowner's jobs, so the required format for the tender can be maintained and the list of jobs to be carried out by the production staff, in a form that permits (relatively) accurate estimates and the correct assignment of man-hours to tasks.

As the contract progresses, the initial estimate, updated as new work emerges or items are deleted, is used to measure progress and develop the actual invoicing costs. Once created, the forecast is potentially a powerful tool for project management, but only if it can be made much more systematic than is usually the case.

The role of IT in commercial ship repairing is historically patchy. Isolated systems have long been used in specific functions. For example, spreadsheets for estimates, planning software for Gantt charts and accounting software for finance are all regularly used. However, often the systems have been developed in an ad hoc way by individuals with some IT skills. As a result, while they are often effective in a limited area of operations, they do not assist data sharing. In addition, the frequent turnover of staff can sometimes result in loss of capability when someone moves on.

However, unlike in the shipbuilding sector, integrated systems covering all yard activities at all contract stages have not been widely adopted. Instead, there have been efforts to use various IT systems, but this has often had mixed results.

Accounting systems provide an accurate cost outcome for a contract, but they are designed for historical cost recording. They can struggle to deal with frequent, ongoing changes to work scope or cancellations. They may not supply information quickly enough or in a suitable format to allow control of day-to-day operations.

ERP (Enterprise Resource Planning) systems are designed for manufacturing operations. These are generally intended to be generic for different industries but often reflect their origins in a particular company. They are based around the use of a pre-developed and optimised plan involving routine, low variation work. Change is generally to be avoided if this is at all possible. Updates require considerable effort, and systems may struggle to keep up with fast-moving contracts. There are large numbers of potential suppliers to ship repairing, and low quantities of materials are usually ordered. There is also usually a frequently varying workforce, and ERP systems are designed to assume that the workforce will be stable. The result for users of ERP systems is an excellent effort to input data with limited output.

Job shop systems are more closely aligned with the service nature of ship repair. They seek to maximise the efficiency of a fixed resource through re-scheduling of job sequences within resource constraints. This suits the industry sector with non-unique products and flexible completion dates (e.g., road vehicle maintenance where another vehicle can be substituted if an unexpected repair delay is encountered) but not ship repair. Ship repairing work uses variable resources to ensure the maintenance of the end date for a contract.

Some systems developed for ship construction have been adapted for repair work, principally for military applications. These may carry a significant overhead in data management and apply in a short, commercial contract. As with ERP, the underlying assumption is that the work is pre-planned, and changes will be limited.

Bespoke systems offer another alternative. Developing a system from scratch allows the specific business processes in a particular ship repairing yard business process to be perfectly accommodated. This also goes a long way to minimising the unavoidable cultural changes encountered when the system is introduced in the yard. However, the costs of a system that is developed from the ground up are high, and in the case of most small and medium-sized repairing shipyards, it may be considered by management to be prohibitive.

The information management system is fundamental to the successful management of a ship repairing contract is. Ideally, the information which is created as an estimate is later developed so that all decisions, actions, purchases, man-hour expenditure and, perhaps most importantly, variations are accurately recorded. Then the information must be made available as close as possible to real-time to selected staff.

The management of a shipyard must analyse their business carefully, consider all the possible alternatives and costs, calculate the benefits of a system, and choose what appears to be the best solution. It is essential to use a demonstration version so that potential users can understand how a system operates and comment on its practicality. Some benchmark tests are also necessary to be confident that the system capacity is adequate for the likely data volume generated in the shipyard. The system should also be able to adapt to any specific shipyard requirements. The shipyard should not need to make significant changes to its operating procedures to fit into the requirements of an IT system.

References

1. Stewart, H.P.: Successful production of a competitive fixed-price ship repair job. Marine Technol. **34**(2), 96–108 (1997). The Society of Naval Architects & Marine Engineers, Jersey City, NJ, USA
2. Evans, M.: Ship Repair Information Management Systems, Ship Repair Technology Conference. Newcastle University (2008)
3. Bruce, G.: Improved Estimating as a Basis for Ship repair Project Control, ICCAS 2011, RINA Conference (2011)

Chapter 12
Human Resources

"Management is nothing more than motivating other people." Lee Iacocca.

To repeat an obvious observation, all ships are built and repaired by the people in the shipyards. So, dealing with the people, the human resources are probably the most critical element in shipbuilding and ship repairing management. Whereas equipment can be purchased at any time, workers must be recruited and often trained. They generally choose where to work and need an incentive to be retained because they are always free to leave for alternative employment. When the shipbuilding and ship repairing industry is busy, there is often a shortage of skilled labour. When there are few orders, shipyards must reduce their workforce, and those who are released may go to other countries, may find new employment in another industry or, if older, may simply retire. Then when orders for ships increase, the workers may no longer be available. (See Bruce, et al. [1])

Marine production generally requires large, complex organisations with a high degree of labour specialisation. Recruiting and retaining sufficient workers with the right skills is a universal problem for ship repairers. Skilled labour must be retained to maintain the business as an operating concern. Although outsourcing is frequently regarded as a solution by ship repairing companies, it often merely delays the problem of obtaining sufficient labour.

The context in which a company needs to develop people will vary. Some relatively new companies are usually set up in areas of low wages and high unemployment where there is a firm intention to create industries as part of a regional or national development programme. There is a significant investment, and such developments are often government-supported. In these cases, there is a clear need to recruit and train a labour force. The recruitment is mainly young workers, usually under thirty on average. Such developments are often on greenfield sites in areas that have not previously been industrial.

Developed companies have typically been in business for twenty years or so. The initial trainees have developed into mid-career, skilled workers, and if there is limited loss of these to other areas or companies, then training and recruitment can be less in demand. If the industrialisation of the site where the company is located progresses, then the loss of skilled workers and recruitment problems may occur. Workers may also be attracted to other countries in the region where a labour shortage exists. The workforce average age is typically mid-thirties to mid-forties.

Mature companies, usually established for decades, may find some of the current workforces are approaching retirement age. These are generally in well-developed countries with diverse industries, including newer sectors that are generally more attractive to younger workers. In these cases, there is an ageing workforce and problems in recruitment and retention. One popular solution is to recruit labour from areas where the ship repairing is newer, wages are generally lower, and a move can be attractive to the workers who have been trained. Typically, the workforce average age is mid-forties to mid-fifties.

Depending on the company context, the needs, and responses in maintaining a skilled workforce will vary. As a starting point for determining human resource needs, it is necessary to establish a business profile under consideration. This will consider several factors.

The company's location, essentially whether it is in a developed area with a large population or is new, will be a significant determinant of the ability to recruit. Also necessary is the maturity of the company and the industrial landscape. Unless the company is relatively new, there will be an ongoing need to replace current staff as they move on or retire.

The product mix is a factor since the more complex the ships to be repaired, the more extensive the range of skills and higher skill levels required of the workers. Also, the technology in use will determine the nature of the skills and the number of workers needed. Whether in an industrially mature or newly developing area, the company's location will be a factor in how a company responds.

The current intake of apprentices and other trainees, skilled workers and graduates or otherwise higher qualified people will help identify the future requirements. In addition, the current employment level of full-time and casual workers and the age profile also help define recruitment needs.

The company profile will be arranged according to the various job titles, categories of skills and succession planning. The qualifications for the entire workforce will need to be established as a starting point for in-service training. (See UK DTI [2])

Having profiled the existing situation, the need is to understand the requirements to deliver the company's output. There is also a need to create a training and development plan to meet the future needs of the business, whether by direct recruitment of qualified people, training, or a mixture of both.

To determine what needs to be done in a company to ensure sufficient skilled people for the future, the starting point is the future company strategy. This will have identified the target ships to be repaired in numbers and types. This, in turn, will allow estimates to be made of the organisation and production requirements from each activity in the company. The current productivity will be the basis for future changes, including any plans to invest in new technology. The location of the shipyard in relation to population centres and sources of skilled workers and potential recruits will also be a factor.

From the analysis, a picture of the current and expected future employment in the company will be created. This will be based on the profile of the company activities, which is what must be done in the company to complete contracts. This will cover all aspects from marketing to planning, production, testing and commercial activities. From this, those activities which are most in need of people can be highlighted. As part of the process, an age profile of the workforce will be created. This will usually look at intervals of five or ten years and identify the activities with the highest need. However, sometimes particular activities can have an older existing workforce or may need more workers.

The ratios of full time to part-time, casual, and sub-contract labour will be identified, and over a period, trends can be monitored, and decisions made. The use of informal labour will depend on availability, the skills required and when those skills are needed. An important consideration is that the formation of a skilled worker takes time. It is possible to develop limited skills quickly, but a fully skilled worker takes several years. Several years' experience is also needed before workers are fully competent and can operate without detailed supervision.

Training existing workers can expand their skills or prepare them for a different role as the product mix changes. Overseas workers can provide a solution to shortages. This is a general problem in ship repairing and shipbuilding, with many countries using workers from neighbouring states. This usually occurs when the area with the shortage can offer better payment rates than the workers' home countries. It can sometimes lead to a chain reaction, where several countries are using their different neighbours' labour forces.

The numbers in each shipyard function are found, typically initially divided into groups. In the first instance, these can be simplified to five, which are:

- Skilled workers who have a specific capability. Accounting and other support staff are also in this category.
- Unskilled workers who are labourers or are semi-skilled with only a limited range of abilities.
- Supervisors, who are very experienced and often well educated, are typically required in a ratio of one to every ten workers.
- Technicians with specific skills in support of design where required for conversions. Also needed are and production staff, welding specialists, test and inspection staff and others.
- Management, including supervisors, who all have specific roles in overseeing the activities of the shipyard.

The workforce has then been classified by age in bands of five or ten years, skill category, and the various activities needed. With this information available, attention can be turned to the future operations of the shipyard.

Once an overall picture of the available labour force has been developed, the shipyard's capacity has to be examined in more detail to determine the capacity for the various activities.

Matching the operational capacity of a production system to demand can be difficult for many reasons. This is the case in the short term for a contract and the longer term when planning the future shipyard strategy. Beyond the immediate order book for a shipyard, which is often a short term for ship repair, the future demand is uncertain, and hence the resource requirements are pending. Furthermore, it is always possible that the types of ships in the forthcoming order book may change, and then the balance between different worker categories will also require to be altered.

Large ships may be in demand, and the absolute number of workers may need to increase. But on the other hand, the numbers required may also decrease with a weaker market, and the number of workers employed will have to be carefully managed to retain the more skilled and experienced. This is a problem because if a potential decrease in the workforce is well known to the workforce, it is precisely those better-skilled workers who will find it easier to move to other employers.

There may also be changes in technology and substantial modifications planned in the facilities and equipment to be used in the shipyard. These changes will always be associated with a plan to increase production or reduce production costs for a consistent output level for an established shipyard. Reductions in worker numbers may again be required, or re-training the existing workforce, potentially destabilise the workers and lead to unexpected losses of skilled people.

For the short term, a specific contract or contracts may require short term changes in the workforce. So, a temporary peak in numbers for a particular category of worker may be needed. It must be decided how best this can be done. Additional hours of work may be used, although this would usually be for a concise period to be acceptable to the workforce. Casual labour might be employed, assuming that adequately skilled people can be found relatively short term. More usually, some of the work will be sub-contracted.

Some repair shipyards have adopted monthly rather than daily or weekly hours for the workforce. A set number of hours per month will be required to be worked, but the precise hours will depend on the production workload. If there is an urgent repair or a large contract, this will result in overtime and weekend working, but the workers will give time off when the workload is lower. This benefits the company and can have some attractions for the workers.

Transaction costs are a potential problem when sub-contractors are used. These are the actual and hidden costs of using those sub-contractors, including insurance, contract costs, risk of failure to deliver and quality assurance. These can apply to any supplier to a shipyard, not only sub-contractors but where a new sub-contractor is used, or the decision is taken at short notice, there will often be no history with the contractor on which to judge the risk. Or the risk may be accepted because of the urgency of a particular situation on a contract.

Measuring the capacity of a shipyard is problematic in ship repairing because of the variety of operations undertaken to bring a ship back to complete operational standards and because the future resource requirements are only an initial estimate when the conditions are determined.

Managing the capacity in terms of the labour force is essential. It is relatively straightforward to determine average demand levels over the next year or so. The average can be calculated using productivity figures from the past, modified by planned changes in technology and products. This information can then be used to determine the required capacity. However, the demand for resources as set by an unmodified programme will not be consistent. There will inevitably be variations leading to peaks and troughs in the number of workers needed. The actual demand will therefore fluctuate, and it is then necessary to deal with the variations.

The requirement for labour in a company throughout a project will increase from the start of work, then rise to a maximum level and decline towards completion. If several projects are in progress at any one time, the overall variation can be reduced, and the work schedule can also help reduce the variation. If the labour force numbers are set for the maximum demand, there is wasted capacity at the start and end of the project. If the capacity is set lower, then there is insufficient capacity, and additional workers need to be found. A compromise level of the workforce can be set to try and minimise the variation. This would try to set the numbers so that any small losses and extra worker costs are kept as low as possible.

In the longer term, if demand rises, the organisation is under-resourced for the duration of the additional demand. If demand falls, then there is under-utilisation of the capacity which is available. The cost of either is high because organisations cannot carry surplus capacity, costing money but not making income. When there are extra resources needed, the use of sub-contracting can also be expensive. As an alternative, the workers can have additional training to be multi-skilled, allowing some labour flexibility. A large organisation may be able to transfer labour between departments or sites.

There are benefits in using sub-contractors rather than having unused but paid for man-hours. The potential benefits of using a sub-contractor are clear where there is a lot of uncertainty about the man-hours that will be required. However, too much reliance on this approach can be dangerous. Overtime working is another option in the short term, but the ability to do this may be limited by workforce attitudes and the costs of a wage premium. So, overtime will potentially be as costly as using a sub-contractor. Of course, the additional cost must be considered in the potential expenses of late shipping a ship back to the owner.

The above emphasises the need to fully understand the available skills, expand these, and know how to do so at short notice.

In the medium term, more pre-planning can be used to sub-contract elements of the work on a contract, subject to the potential cost implications as outlined above. There is also more time to carry out re-training of some of the labour force so that flexibility is available.

In the long term, it should be possible to re-structure the organisation, as part of the shipyard strategy development, so that the balance of labour between different activities and skills is changed to suit, for example, a changing product range. It is also possible in the longer term to adopt more automation, replacing workers with machines, although the machines do have a fixed cost. The basis for this and other options is a thorough review of the labour force as described above. However, realistically, a large proportion of the repair work will always be completed onboard the ship, where any automation is always problematic.

After completing a review of the demand for resources, their availability must be considered. The resources will vary over time; for example, the production equipment will often deteriorate over time and replacing equipment is a normal part of the business. The labour force will also get older, and apprentices or other recruits may not be available. Some skilled labour may leave for better conditions or can't be re-recruited after they have been laid off. To some extent, if a consistent product mix is maintained, some labour losses can be balanced through learning curve effects and general performance improvement.

Earlier, the problems of recruiting shipyard labour in advanced nations were mentioned, and this is, in fact, an increasing problem in many places. When a country is industrialising, the classic movement is first into heavy industry, then lighter industries such as consumer goods and services or electronics. Although that is a grossly simplified description, it broadly operates in most places. In general, people prefer to choose comfortable working conditions. (See Hart and Schotte [3])

The result for a developing nation is an initial move from agriculture to heavy industry, giving workers a more stable and usually increased income. The heavy industry is also initially able to absorb a lot of relatively low skilled labour. The work is labour intensive, often heavy manual labour, and is usually in the open air is not of great concern to a population used to agriculture. Such a startup industry is not very efficient, and as labour costs rise, there is a need for more extraordinary skills so the sector can become or remain competitive. There will also be a need for a proportion of skilled labour to be recruited from the startup for technically complex work, along with significant reliance on equipment suppliers and sub-contractors.

If industrialisation is successful, then lighter industries will emerge. These operate in enclosed factories where the working conditions are better than in a shipyard, where some of the work is conducted in the open air, in confined spaces and at height. The factory work is also less physical, so more attractive to workers. There will also be increasing numbers of jobs in transport and services, again potentially more attractive. Consumer goods are likely to follow, with factories set up to cater for increasing local demand. Also, large, often multi-national companies will be drawn to set up new factories, which can take advantage of a labour force increasingly used to industry and still accept relatively low wages. At this point, the ship repairing industry is likely to find increasing difficulty in recruiting and retaining skilled workers.

Finally, in a well-developed economy, increasing demand is found for electronics, luxury goods and services. Often there is still a lower comparative labour cost to attract international companies. Also, increasing affluence goes alongside improved education standards, leading to a demand for higher-skilled and higher paid work.

As a result of the changes described, the ship repairing industry may have a limited lifespan in any country. Arguably, as countries change more rapidly and move to newer industries more quickly, the lifespan is shortening. If the location is close to terminals or passing ships, there will almost always be local demand for repair. However, there is no guarantee that an operation to take advantage of this will be profitable. Where ships are passing or calling at a port and established docks and other facilities, it is often the case that companies fail, and a new company then takes on the shipyard. This new operator can start with low overheads and recruit only when required for specific contracts, subject to the dangers outlined above.

Other social issues affecting ship repairing include health and safety legislation and trade unions. Repair of ships was always a hazardous industry. In the nineteenth century and into the twentieth century, safety was almost non-existent. Several deaths during a year of ship repairing operations were regarded almost as an acceptable occupational hazard. Legislation to promote safety was introduced mainly during the latter half of the twentieth century. Occasional deaths in advanced companies still occurred but always became a focus for an enquiry, improved safety procedures and often new legislation.

The effect has increased costs, which can be a crucial factor in moving to newer countries. Visiting shipyards in the late twentieth century, there was a strong contrast in many business areas. Scaffolding to provide access to high working areas is a good example. As an example, specialised metal scaffolding frameworks were used in the long-established shipyards in Europe and Japan, usually with safety fastenings. Metal bound wood planks were used for the working platforms, with toe boards to prevent slipping off the platforms. Handrails were a feature, with purpose constructed stair-cases and sometimes elevators. In any particularly hazardous areas, safety harnesses were attached to the workers to prevent falls. Powered, mobile access equipment is common to replace the traditional scaffolding with a safer alternative largely.

In contrast, scaffolding was often made of bamboo or other wooden poles lashed together with rope in new shipyards in developing countries. Flimsy work platforms with no toe boards were provided. Instead of fixed ladders or elevators for huge ships, workers simply climbed the scaffolding to reach their work areas. This resulted in a much lower cost operation, but at the expense of worker safety.

However, the faster pace of change and international pressures spread the need for safety around the world more rapidly.

Occupational health is also an issue that has become prominent in recent years. The dangers of hazardous materials, noisy equipment, vibration, and noxious gases have been identified, and action is taken to avoid them. Until the late twentieth century, many processes in everyday use have been phased out and replaced by quieter alternatives.

When ships were still riveted, hammers and later pneumatic hammers were used, in large numbers, producing a noise level more than enough to cause deafness in later life for workers using the equipment and those in the same spaces. In the mid-1960s and early 70s, pneumatic equipment was still used for caulking welded seams, producing the same effects. A single piece of equipment in a confined space could make conversation impossible within twenty metres of the noise source.

The introduction of welding to the industry on a large scale resulted in a poisonous atmosphere where the process was used. Again, confined spaces produced the worst effects, and the operators and those near them were exposed to toxic fumes. Over time, masks and sometimes breathing apparatus were introduced to minimise the hazards. Eye damage was another problem. Although welders were supplied with darkened eye shields built into a helmet, others in the area could have issues with looking directly at a welding arc. Improvements in welding and the use of shielded arcs have improved the situation. The elimination of some facilities, such as blacksmiths workshops with open coke fires and red-hot metal, has also enhanced the enhanced part of the working environment in shipyards. Vibrations from pneumatic tools could also cause muscle and nerve damage to hands and wrists after prolonged use and were still in use.

Personal protective equipment is mandatory in most shipyards. This includes hard hats, safety glasses, ear defenders because a shipyard is still a relatively noisy environment, gloves, steel soled and capped work boots, and company supplied overalls. However, small, developing shipyards often had workers in regular shoes, even sandals, with no other protective equipment provided.

Although the modern shipyard is a far better working environment than it was historically, it is still not a place many people wish to work in.

There is a further question about payment systems for an existing shipyard or a newly recruited development. There are basically two alternatives available: payment for hours worked and secondly payment by results. In the first case, there are set working hours and an hourly pay rate, so workers are paid for attendance. There may be fines for late arrival, and typically the wages for half an hour are deducted. This is a sound system in a well-managed shipyard as the work will be planned, and materials will be available for production, so the workers stay busy. However, if the organisation is not so good, workers may be waiting for some requirements and not contributing to production. This creates a severe problem for the management and could consider the alternative of payment by results.

In general, ship repairing companies hold daily or twice daily meetings to schedule immediate work. The tasks are assigned to shipyard workers, or a sub-contractor, with a deadline for completion. Results pay sub-contractors, that is, completing the assigned task.

This payment by results option has been available to management in all sorts of industries, and agriculture, for as long as these have existed. In this case, the workers are paid for what they can produce. So, in a shipyard, payment could be for the number of items made, the metres of welding completed or any other readily measurable output. The attraction is that this payment approach should benefit both parties in that management gains more results and workers have higher pay.

This approach has grown recently with the increasing international competition forcing managements to seek cost reductions. Along with outsourcing, trying to obtain more production for less cost is a driver for many shipbuilders. However, the result is not always so beneficial. Sometimes the workers will be content with a level of pay and restrict output, and at other times if the management cannot supply enough work, there will be labour trouble. Management may want to reduce the high pay and set minimum output targets that the workers regard as too difficult. In pursuit of increased pay, the workforce may produce lower quality output than is required. Alternatively, the management may wish to restrict payments by rejecting some of the production not always fairly.

When it works effectively, payment by results can be beneficial, but often it is a source of friction between management and the workforce. For example, the workers do not generally control the supply of materials and interim products. If there is a shortage of these, then no work can be possible, and the management will not pay. The workers may decide informally to limit their output so that the wages are adequate but will not choose to produce more for a higher payment. This is most likely towards the completion of a shipbuilding contract. The workers will slow down, hoping that the work will be available for a more extended period if they slow down the pace. They may consider it possible that the management will offer a higher rate to ensure the completion of a ship on time.

There is an alternative option to pay workers for hours worked during a longer period in many ship repairing companies, usually a month. The total hours are equal to the conventional daily hours but can be used at the discretion of the management. If there is a high workload, the workers will be required to work overtime or at weekends. When the workload reduces, the workers can take time off. This allows the management to avoid overtime payments or the use of casual or subcontract labour.

There is increasing international concern about environmental pollution. Concern about the health and safety of workers in shipyards is mirrored by concern for the public, especially those living or working close to the industry. As a result, there is increasing legislation to control or ban some industrial activities which cause pollution. Several shipyard processes can cause problems. These include surface cleaning, shot or grit blasting of steel, paint coating of steel, waste disposal in general, welding, especially where fumes are released and noise.

Environmental legislation is found in most countries, although the levels of enforcement may vary considerably. In most, to carry out an industrial activity that generates wastes, it is necessary to obtain a licence from the appropriate regulator or inspectorate. Most environmental management agencies will use a mixture of enforcement to challenge polluting behaviour and assistance to companies to avoid pollution. Some can advise on how to improve operations to prevent any adverse impacts on the environment. Sensibly preventing an incident is far better than handing out a fine.

Assistance in interpreting and conforming to legislation is also usually available for the same reasons. A similar approach is generally used for quality management. The relevant standard is ISO 14000, and this is like ISO 9000 for quality management in that the focus is on the production processes, which will generate no or minimum waste. As with ISO 9000, certification is carried out by accredited third-party organisations rather than by ISO directly. Again, a similar approach is used to review the processes, with wastes rather than products as the outputs. (See Bruce and Newell [4])

Establishing ISO 14000 in a company is a project. First, there is a requirement to review the company and deal with wastes and other environmental problems. Then it is necessary to implement the standards, generally with a third party, accredited advisor, as with quality assurance. Then, once the system is in place, the requirement is to monitor how it works and the results obtained and adjust and improve if required.

Environmental management has a fundamental international objective: to ensure that pollution is minimised or avoided altogether. Although the emphasis is on managing waste or other pollutants, not necessarily on the processes to be used, many industrial processes may not be used if there is no environmental management plan in place. The agency usually has the power to order work to be stopped if the process is not complying with requirements.

Often, companies are required to monitor their performance and are responsible for compliance with the legislation. There are inspections from time to time to ensure compliance in most cases. However, the inspections are often as much about record-keeping, rather than checking any actual pollution, which might be considered a system to blame in the case of unexpected problems, rather than avoiding the problems in the first place. As a result, the information on waste management is often more used for investigations after the event than for active avoidance of pollution. However, for shipbuilders, there is no alternative to compliance.

A problem for ship repairing, which it shares with other large-scale industries, is that the site is large, the activity level is high, and any problems are readily observed. It is easier to check on a shipyard than on perhaps one hundred small businesses where non-compliance with legislation is easier to hide. These produce smaller individual quantities of waste, and so it is easier to conceal any pollution.

The companies should use appropriate technology in their operations and must still comply with regulations. Overall, the regulations and their enforcement can be expected to become more restrictive. Often the legislation is not prescriptive companies may operate in any manner, using equipment that is not disallowed for health or safety reasons, provided they do not pollute. Self-monitoring is the expected practice simply because the regulators cannot monitor the industry continuously.

Noise is one form of pollution that can affect local communities, and local authorities often carry out enforcement. Noise pollution is primarily a health and safety issue but also affects the local environment. It is a significant problem at night when people are trying to sleep, and the level of ambient noise is low because there is no traffic on roads. It is the most common cause of complaints, especially from local populations, about any industry. Depending on local and national regulations, receiving many complaints can be problematic for shipyard management.

Noise sources include services, with generators, compressors, pumps, and frequency converters as a large part of the problem. However, services are also a source of leakages, and noise can be generated at end-use consumption.

Noise can be controlled by choice of quieter processes, probably the prominent example in the past being the elimination of riveting which was responsible for large scale deafness among shipyard workers. More recently, the elimination of pneumatic hammers has also reduced noise levels. In addition, management of the noisy processes in use can reduce the effects locally by soundproofing and, as a last resort restricting operations to daytime when the problem is less noticeable.

Surface cleaning and blasting are necessary to prepare steel surfaces for over-coating. A range of processes is available, including mechanical methods. These can either be portable machines, sanding, grinding and even small hand tools fall into this category. Grit blasting contained in a small, moving cabinet is also limited to small areas of repair.

Blasting is often necessarily carried out in the open air, which is an immediate potential problem, especially if the shipyard is close to any population. One solution is to use water injection to prevent airborne pollution. High-pressure water blasting for washing and ultra-high-pressure water blasting for surface preparation to avoid airborne pollution are also normally used for ship repairing. Dry blasting with grit, which is then recycled, is the preferred method for steel cleaning.

Some use has been made of recycled glass for steel blasting. This has been mainly on a trial basis. The process produces a good standard of surface, and importantly the waste material is dust. The residues of glass are harmless, but they must be separated from the paint and other debris. Generally, the waste from cleaning causes a significant disposal problem.

The preferred application option for coatings is an airless spray. It is fast, provides a good quality of painting and is safe. The work must be completed in the open air or the building dock. Outside, the problems are associated with overspray, where windy conditions increase, and paint particles carry long distances. Operator training is essential to manage the quality and to minimise the wastage.

Welding is primarily a health and safety issue. Avoiding fumes close to the workplace is essential by means of extraction systems. These can be within workshops, in work locations on the ship, or integral with the welding equipment. Air-fed helmets are used for confined spaces, and over time, it may be more possible to replace people with more automation or use remote control where possible.

Waste disposal is essential to maintain a clean and tidy workspace to avoid hazards and manage the large quantities generated by ship repair. Reasons to reduce waste include avoiding the high disposal costs in landfills or by incineration. There are increasingly strict hazardous waste regulations, and generators of hazardous wastes are liable for their waste even after it is transported somewhere else for disposal. Waste reduction is perceived favourably by the community and can enhance public image. However, waste is also a risk factor if anything does go wrong.

A company should encourage employees to assist in identifying wastes. This includes leftover materials, use of energy and outputs, including products and waste. A review of what is being disposed of or released into the environment is required. There are options to reduce waste, reduce the sources by changing materials where feasible, choose materials that can be managed on-site, improve the processes in use and improve the housekeeping on site.

They should also decide to contain any spillages, use offcuts of pipes and steel, inspect the site for leaks in services and remove any waste left around. It is also possible to train employees in better materials handling to avoid breakages and maintain materials in protective environments. Immediate collection of any waste can avoid hazardous waste contaminating large quantities of other materials which might be useable or saleable. Mixed wastes are more difficult to treat and/or dispose of because contamination may make otherwise reusable waste useless.

It is essential to reuse any waste where this is possible and otherwise to recycle waste. Often a specialist sub-contractor is used to dispose of the wastes. The workers are encouraged to sort waste at the source into containers which can then be regularly collected. Different materials are kept separate so they can first be reused if possible. An example is shot expensive blasting materials. The used material is filtered to remove debris and then sorted by size so that a mix of new material and clean, partly used material is used. This gives very effective cleaning. The material and debris which are not reusable are then sent for disposal.

The same objectives of managing waste apply to all other materials at all stages of production for all activities. However, it can be not easy if work-study is applied. The value of careful sorting and reuse or disposal can be decided, and there is a potential benefit to the shipyard.

It is important to remember that if environmental management is carefully carried out, there is a potential benefit in reduced costs, workplace safety, and, of course, a better workplace in the immediate location.

References

1. Bruce, G., Hills, W., Granger, N.: The Skills Crisis in UK Shipbuilding and Repairing (with Hills and Granger, RINA North East Joint Branch, Nov 1998, RINA Transactions 1999 (1999)
2. Skill Requirements in the Shipbuilding and Ship Repairing Industries, Joint UK DTI/SSA report sponsored by UK DTI (1997)
3. Hart, Pt'., Schotte, D.: European Shipbuilding Social Dialogue Committee. HR Research Study: Demographic Change & Skills Requirements in the European Shipbuilding & Ship Repair Industry (2008)
4. Bruce, G., Newell, P.: The economics of implementing ISO 14001 in a shipyard. SNAME J. Ship Prod. 17(3), 135–144 (2001)

Printed in the United States
by Baker & Taylor Publisher Services